THE SCIEN
MONSTERS

The Truth about Zombies, Witches, Werewolves, Vampires, and Other Legendary Creatures

MEG HAFDAHL & KELLY FLORENCE

Skyhorse Publishing

Skyhorse Publishing books may be purchased in bulk at special discounts for sales promotion, corporate gifts, fund-raising, or educational purposes. Special editions can also be created to specifications. For details, contact the Special Sales Department, Skyhorse Publishing, 307 West 36th Street, 11th Floor, New York, NY 10018 or info@skyhorsepublishing.com.

Skyhorse® and Skyhorse Publishing® are registered trademarks of Skyhorse Publishing, Inc.®, a Delaware corporation.

Visit our website at www.skyhorsepublishing.com.

10 9 8 7 6 5 4 3 2 1

Library of Congress Cataloging-in-Publication Data is available on file.

Cover design by Peter Donahue
Cover photograph by gettyimages

Print ISBN: 978-1-5107-4715-9
Ebook ISBN: 978-1-5107-4718-0

Printed in the United States of America

We dedicate this book to our parents, who let us watch horror movies.

CONTENTS

INTRODUCTION

I t was nearly twenty years ago when we met because of a T-shirt. I (Meg) was working in a gift shop when Kelly strolled in, *The X-Files* emblazoned across her chest. An immediate connection was made, the special kind that forms when two people are thrilled to share in their fandom. Before that day, I had spent countless afternoons screening *The Shining* for my friends, none of whom cared for it or understood my horror obsession. As our friendship grew, Kelly and I eagerly shared movies with each other, from the silly, low-budget *Invasion of the Blood Farmers* to the truly terrifying *The Ring*. (We spent that night wide awake, convinced every gust of wind was Samara preparing to crawl inside and eat us whole!) Our affection for horror became a touchstone in our lives, a catalyst that has brought us to create, consume, and better comprehend the complexities of the horror genre.

This book is a fun and approachable way to understand the science behind the monsters of our favorite horror films, the ones we came back to again and again, whether we were renting from Blockbuster, or watching our own scratchy VHS copies. We've devoted our lives to the study of film, literature, communication, and true crime, and so we relished this chance to delve deeper into the world of science and to then connect it to the world of horror. We had the opportunity to conduct research and visit with experts at the top of their fields; from scientists at the Mayo Clinic to creatives such as Simon Barrett, the screenwriter behind *Blair Witch* and *V/H/S*. Come with us on this journey to unveil the thrilling and sometimes disturbing truths behind the creation of horror's greatest monsters.

SECTION ONE
SLASHERS

CHAPTER ONE

HALLOWEEN

Year of Release: 1978	
Director: John Carpenter	
Writer: John Carpenter, Debra Hill	
Starring: Jamie Lee Curtis, Donald Pleasence	
Budget: $300,000	
Box Office: $60 million	

L ike most children growing up in the 1980s, we saw many horror movies on VHS for the first time. The nostalgia of visiting a video store, perusing the covers, and choosing the weekend's haul holds countless fond memories for us. (In fact, it's the experience that inspired us to create our *Horror Rewind* podcast.) *Halloween* was no exception. The plot of this 1978 horror movie begins with a six-year-old boy named Michael Myers. Dressed as a clown, Michael murders his sister by stabbing her with a knife that looks terrifyingly large in his tiny hands. This shock of a child, ready for trick-or-treating, brutally subverting our expectations of innocence is at the heart of *Halloween*'s appeal. Actor and comedian Jody Kujawa reflects on his experience seeing the movie for the first time:

> That movie sat with me more than most horror films. It seemed like the most simplistic plot: a guy just wants to kill a lady for no apparent reason. But there is no greater horror than that. The thought that someone is out there, watching us, and they have decided we are going to die. Jamie Lee Curtis also really delivered a personality to the screen that later slasher knockoffs would never be able to

correctly mimic because they missed the subtleties of Laurie Strode that made us terrified that she would die. Jamie Lee Curtis and John Carpenter are horror film royalty.

Jamie Lee Curtis was just nineteen years old and an unknown when cast in *Halloween*. Most of her dialogue was written by producer and co-writer Debra Hill, who had been a babysitter herself in her youth, and prided herself in telling women's stories from women's perspectives. *Halloween* not only helped solidify Curtis's status as a scream queen but also popularized the final girl trope and slasher films in general.

Was *Halloween* based on a real person? There is a claim that Myers could be based on Stanley Stiers, who was said to have gone on a killing spree in Iowa in the 1920s. He allegedly murdered his entire family on Halloween. Although the details of this urban legend are strikingly similar to the plot of *Halloween,* and the story itself is heavily shared on fan sites, there are no credible sources for this event.

John Carpenter, the writer and director of the cult classic, recounts being inspired to write the film while visiting a mental hospital for a class in college.[1] "We visited the most serious, mentally ill patients. And there was this kid, he must have been twelve or thirteen and he literally had this look." The look is described by the lines Carpenter wrote for Donald Pleasence, who played psychiatrist Dr. Sam Loomis: "This blank, pale emotionless face. Blackest eyes. The devil's eyes. I spent eight years trying to reach him and then another seven trying to keep him locked up, because I realized what was living behind that boy's eyes was purely and simply evil."

Some people look at the character of Michael Myers and see someone who embodies everyone; a part of themselves that could someday snap. Are we born evil? Are we all capable of murder? That brings us to the plausibility of the plot. Are children capable of murder? According to an article in *The Atlantic,*[2] psychopaths are among us. Children with psychopathic tendencies are described as having "callous and unemotional traits." This includes characteristics and behaviors such as a lack of empathy, remorse or guilt, shallow emotions, aggression or even cruelty, and a seeming indifference to punishment. Researchers believe that nearly 1 percent of children exhibit these traits. In 2013, the American Psychiatric Association included callous and unemotional traits in its diagnostic manual, DSM-5.

Studies have found that kids with callous and unemotional traits are more likely than other kids to become criminals or display aggressive, psychopathic behaviors later in life. While adult psychopaths constitute only a tiny fraction of the general population, studies suggest that they commit half of all violent crimes. Researchers believe that two paths can lead to psychopathy: one dominated by nature, the other by nurture. For some children, their environment can turn them into violent people with a lack of empathy. Those who grow up in abusive homes, or are neglected, may show more traits in common with those who are diagnosed as psychopaths. For other children, a loving home environment doesn't prevent them from displaying the traits.[3]

What are some warning signs that a child could be a potential murderer? The biggest red flags may be an affinity toward violence and a lack of feeling or recognizing others' feelings. According to "My Child, the Murderer,"[4] parents of killers recall their children getting into trouble in school more often, being bullied, or withdrawing from others.

Are there many instances of children murdering others? There are numerous cases of murder being committed by children over the course of history. Some notable child murderers include Mary Bell who committed the first of two shocking murders on the day before her eleventh birthday. In May of 1968, Bell and a friend strangled a four-year-old boy. A month later, and joined by that same friend, Bell strangled a three-year-old boy in the same area as the first killing. She returned to the body and carved an "M" into the boy's stomach, along with scratching his legs and mutilating his genitals. Bell was convicted of manslaughter and released in 1980.

In 2000, a six-year-old boy named Dedrick Darnell Owens killed a classmate in Michigan. He had previous behavioral issues before the murder, including hitting, pinching, and even stabbing another student with a pencil. After fatally shooting a girl in his class with a gun he brought from home he was released to live with relatives. In an 1893 ruling,[5] the US Supreme Court declared that "children under the age of seven years could not be guilty of felony, or punished for any capital offense, for within that age the child is conclusively presumed incapable of committing a crime."

In February of 2009, eleven-year-old Jordan Brown murdered his father's fiancée, Kenzie Houk, who was eight months pregnant at the time. While the soon-to-be mother was sleeping in her bed in their Pennsylvania home, Brown shot her in the back of the head. Initially, Brown was to be tried as an adult, but was eventually found guilty of first-degree murder as a juvenile.

In an attempted murder case, two Wisconsin girls lured a friend out into the woods in 2014 with plans to murder her. They claimed that they were trying to impress the fictional character Slender Man. The victim was able to survive her nineteen stab wounds but the case has led the public to question whether adolescents should be charged as adults in circumstances like these.

What can we deduce about the fictional Michael Myers? Was it nature or nurture that drove him to kill? As John Carpenter's dialogue revealed, Michael most definitely had a lack of empathy and emotional connection to the world around him. We can assume he displayed some of the other telltale signs that he was a possible sociopath, but could his home environment have contributed to his behavior? Studies show that many children who kill have several things in common; including an abusive home life, isolation from their peers, and an inability to cope. Stefan Hutchinson in his book *Halloween: Nightdance* [6] suggests that Michael Myers's hometown of Haddonfield is the cause of his behavior. He calls Myers a "product of normal suburbia—all the repressed emotion of fake Norman Rockwell smiles." We may never know Michael's true motivation for the murder of his sister, but it won't stop fans from trying to figure it out.

If you've seen the numerous sequels to the *Halloween* movies you know that Michael Myers somehow survives his wounds and goes on to kill another day (except for the third installment of the *Halloween* franchise, but that's another story!). This led us to question if someone can really survive multiple gunshot wounds. Just like the young girl in Wisconsin who heroically survived nineteen stab wounds, there have been many cases of people living after being shot numerous times. Doctors who have treated gunshot victims say that people can survive gunshot wounds as long as major organs such as the heart, brain, and blood vessels are avoided.

In *Halloween* Dr. Loomis shoots Michael Myers and it seems as though the monster is slayed, but as a popular horror trope, Michael cannot be brought down so easily. He is able to get up and continue on his killing spree. Can some people handle pain better than others? There are several factors that could account for a person's ability to work through pain or a major injury. The first is adrenaline. When in a stressful or intense situation our primitive brain alerts our sympathetic nervous system and releases adrenaline.

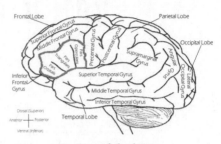

A view of the brain.

An adrenaline rush, or fight-or-flight response, can cause people to temporarily not feel any pain and even allows some people to have almost superhuman strength.

Another possible reason Michael Myers may be able to continue on after being shot is because of a condition known as congenital insensitivity to pain. People with this disorder have an indifference to or are unable to feel pain. This can be very dangerous and for many children it has proven to be fatal. Being unable to feel injuries, or be aware of illness, can cause many symptoms to go unnoticed like burns or even broken bones. Another explanation for insensitivity to pain was explored in popular TV show, *The X-Files* (1993–2018). In the episode "Home," a family who has practiced inbreeding for generations has lost their ability to feel, making them seemingly unstoppable. Others believe Michael Myers is somehow immortal, perhaps for supernatural reasons, and is incapable of dying. Maybe we'll find out in the next sequel. Whatever you believe about Michael Myers's relentless pursuit to murder, one thing is clear: *Halloween* and its impact on slasher movies and culture are solidified for years to come.

CHAPTER TWO

CHILD'S PLAY

Year of Release: 1988
Director: Tom Holland
Writer: Tom Holland, Don Mancini
Starring: Catherine Hicks, Chris Sarandon
Budget: $9 million
Box Office: $44 million

Everyone has seen a doll. Whether it was your own, your sibling's, or a well-loved doll at your daycare, we have all been exposed to their glassy, watchful eyes. Much like children, dolls hold a particular innocence which should make us feel calm and safe. Right now, you might be remembering a good friend who went on adventures with you. A Barbie with tangled, blonde hair perhaps, or a porcelain baby doll you had to rock carefully. As ubiquitous and innocent as they may be, dolls also hold an inherent conflict in their soft insides. They are at once an object and a person. We lock them away in closets and boxes, only to later treat them like friends. It is no wonder that dolls, strangely human, strike fear in our collective hearts.

Filmmakers Tom Holland and Don Mancini capitalized on this fear of dolls when they created the movie *Child's Play* (1988). The story of a killer, Charles Lee Ray (Brad Dourif), transferring his evil soul into the body of a Good Guys doll, resonated with audiences and sparked a lifetime of sequels. There is just something about seeing that small, plastic hand wrapped around a knife that both scares and delights fans of *Child's Play*, but Chucky, the red-haired and freckled doll from *Child's Play* isn't the first haunted doll to terrify people. The idea of haunted or possessed dolls goes all the way back to ancient Egypt. Enemies of Ramesses III were said to have created waxen figures of him, believing his spirit would inhabit the doll. They hoped this strategy could be used to kill him. This is similar to the concept of voodoo dolls or poppets, which could be used to carry a curse or to protect a person. There are some famous allegedly haunted dolls in history, including the Raggedy Ann doll who inspired the *Annabelle* (2014) movies and a Barbie doll who is said to have supernatural powers.

Child's Play took inspiration from the story of a haunted doll named Robert. The owner claimed that the doll was possessed, through voodoo, with the soul of someone intended on torturing his family. The doll itself was based on the popular Cabbage Patch Kids and My Buddy dolls of the 1980s. Through this plot the writers wanted to explore the effects of advertising and television on children. Ironically, members of the public protested the initial release of *Child's Play* in fear that the film would incite violence in children.

Does media inspire people to become violent? In 1982 the movie *Halloween II* inspired a man to stab an elderly couple a combined forty-three times.[1] The perpetrator, while admitting the crime, blamed his actions on a drug-induced flashback to a stabbing scene in the horror movie. *Child's Play 3* (1991) was cited as the inspiration for two murders[2] in the United Kingdom in 1992 and 1993. The perpetrators in the latter became the youngest convicted murderers of the twentieth century.

Prior to the murders inspired by *Child's Play 3* there were already laws in place to try to protect the public from the perceived danger of media. "Video nasties" in the UK were a list of essentially banned or censored films in order to spare the public from excessive violence. The Video Recordings Act of 1984 put stricter censorship requirements on movies being released on video than in cinemas. Banned films included *The Evil Dead* (1981) starring Bruce Campbell and the Italian horror movie *Cannibal Holocaust* (1980). Only decades later were films like *The Exorcist* (1973) and *The Texas Chainsaw Massacre* (1974) finally able to be released uncut in the UK.

Horror film protests didn't just happen overseas. Groups protested many movies in the US including *Silent Night, Deadly Night* (1984). The story followed a boy who witnessed his parents' murder at the hands of someone dressed as Santa Claus. The character grows up to go on his own killing spree at Christmas time. Concerned parents and critics were afraid of children seeing Santa Claus portrayed in such a violent light. Protesters thought the movie would traumatize children and undermine their traditional trust in the mythical figure. Advertising for the film was stopped six days prior to its release and the movie itself had a shortened run due to public outrage.

The Chucky doll in *Child's Play* may not have been delivered by an evil Santa as a Christmas present, but was possessed by an evil person in the film. The character of Charles Lee Ray was a conglomeration of three famous killers: Charles Manson, Lee Harvey Oswald, and James Earl Ray. Charles Manson was a serial killer and cult leader who gained notoriety in the 1960s. Lee Harvey Oswald was the accused killer of President John F. Kennedy, and James Earl Ray was the convicted assassin of Martin Luther King Jr. The character of Charles Lee Ray's past is explored a bit throughout the *Child's Play* franchise and it's revealed that he, too, is a killer.

While the science on possessed dolls isn't readily available, we wanted to know how detectives would determine if a Chucky doll could be the killer at a crime scene. We spoke to Timothy Koivunen, Chief of Police and former detective in Eveleth, Minnesota to find out more:

Kelly: **"What are some things you look for when you arrive at a crime scene?"**

Police Chief Koivunen: "Some of the first things we consider on any call or complaint is to protect the scene at all costs. We will block off a perimeter usually larger at first as we can always lessen or make it smaller. Before entering the scene, consider who may have already entered the scene (witnesses, the reporting party or caller, ambulance, fire, police, family, friends, bystanders, or simply nosey people). If we determine it a crime scene, a log sheet is always started for those who enter and exit as well as who they are and the times. Foot or shoe prints may have to be taken and documented. This is simply just to enter or approach the scene.

The Bureau of Criminal Apprehension (BCA) may be called, and we would protect the scene or assist the BCA. We are constantly scanning and looking for anything out of the ordinary such as blood, signs of struggle, broken or misplaced items, weapons of any kind, or any other evidence that may be pertinent. Then, we would also attempt to locate and interview any and all witnesses, suspects, neighbors, or anyone who may have heard, seen, or have knowledge of the victim(s)."

Meg: **"What evidence would you use to determine a criminal's height?"**

Police Chief Koivunen: "Besides blood spatter to determine the height of a suspect, we may use the location and angle of the wounds or trauma of the victim and maybe defensive wounds on the victim. Also, any witnesses obviously could shed some potential height of the suspect or any surveillance video that may be obtained. We currently have a missing female from a few years ago and had the BCA lab come in and luminal the entire house where she was staying to see if there was any blood evidence. Thankfully, nothing was found."

Kelly: **"How do you think crime scene analysis has changed over the past decade? Where do you see it improving in the future?"**

Police Chief Koivunen: "I think crime scene analysis has evolved greatly over the past few decades with the increase of technology and particularly DNA evidence, fingerprint evidence, etc. Some of the things seen on CSI are simply not reality. We will send in suspect fingerprints that may take three to six months to get a result back. It's not instantaneous like on television."

Fingerprints can help determine a suspect.

As Police Chief Koivunen mentioned, another aspect of crime scene investigation is bloodstain pattern analysis, or the interpretation of bloodstains at a crime scene. The purpose of analyzing bloodstains is to recreate the actions that caused the bloodshed. Experts examine the size, shape, distribution, and location of the bloodstains to form opinions about what did or did not happen. What are some of the things that can be determined? Using geometry and the science of how blood behaves, detectives can discover where the blood came from, what caused the wounds, from what direction the victim was wounded, how the victim and perpetrator were positioned, the movements made after the bloodshed, and how many potential perpetrators were present. In theory, a bloodstain pattern analysis expert could see one of Chucky's crime scenes and determine that the killer was approximately twenty-nine inches in height. They wouldn't be able to tell how many cheeky one-liners Chucky used before the crime was committed, but maybe science in the future can help with that!

CHAPTER THREE

A NIGHTMARE ON ELM STREET

Year of Release: 1984	
Director: Wes Craven	
Writer: Wes Craven	
Starring: Heather Langenkamp, Robert Englund	
Budget: $1.8 million	
Box Office: $25.5 million	

A mysterious syndrome is said to have been the inspiration for the movie *A Nightmare on Elm Street* (1984). According to a 2014 interview with *Vulture*, writer and director Wes Craven said:

I'd read an article in the *L.A. Times* about a family who had escaped the Killing Fields in Cambodia and managed to get to the US. Things were fine, and then suddenly the young son was having very disturbing nightmares. He told his parents he was afraid that if he slept, the thing chasing him would get him, so he tried to stay awake for days at a time. When he finally fell asleep, his parents thought this crisis was over. Then they heard screams in the middle of the night. By the time they got to him, he was dead. He died in the middle of a nightmare. Here was a youngster having a vision of a horror that everyone older was denying. That became the central line of *A Nightmare on Elm Street*.[1]

Is there a medical explanation for what happened to the young man? Sudden Unexplained Nocturnal Death Syndrome (SUNDS) is a condition that strikes men of South Asian descent more prevalently than other demographics, as was described in the story Wes Craven read about. Specifically, there were a startling number of cases of SUNDS among Hmong immigrants in the early 1980s:

> In 1981, the Centers for Disease Control began tracking a mysterious rash of sudden unexplained nocturnal deaths occurring in apparently healthy, male immigrants from Vietnam, Laos and Cambodia. The problem, unknown in other ethnic groups, has now claimed more than one hundred and four men, averaging thirty-three years of age, and one woman, according to Dr. Gib Parrish, a CDC medical epidemiologist. Ninety-eight percent of the deaths occurred between ten p.m. and eight a.m. In 1981, the peak year of these deaths, twenty-six men, often Hmong refugees from the highlands of northern Laos, died in their sleep. Usually victims were simply found dead, but when medics arrived quickly, the men's hearts were fibrillating or contracting wildly, a symptom Parrish said may result from numerous possible causes.[2]

What cultural beliefs could have contributed to this phenomenon? In the Hmong culture, the spiritual realm is highly influential and dictates what happens in the physical world. The spirits of deceased ancestors are thought to influence the welfare and health of the living. The Hmong immigrants may have been experiencing a sense of guilt in fleeing their homeland coupled with extreme stress. In *Sleep Paralysis: Night-mares, Nocebos, and the Mind-Body Connection*,[3] Professor Shelley Adler comes to this conclusion: "In a sense, the Hmong were killed by their beliefs in the spirit world, even if the mechanism of their deaths was likely an obscure genetic cardiac arrhythmia that is prevalent in Southeast Asia." To understand more about the Hmong culture, we spoke to Mai Vang, creator and chair of The Hmong Museum in St. Paul, Minnesota:

Kelly: **"Can you tell us about the Hmong immigration journey?"**

Mai Vang: "The Hmong are an ethnic minority who are from Southeast Asia including the countries of Laos, Thailand, and Vietnam. There are also Hmong who live in southern China. Today, the Hmong people are in almost every country including the United States. The Hmong became involved in the United States' war in Vietnam which started in 1955. By the 1960s the American CIA were training Hmong who lived in Laos how to be soldiers. Although the war was between the US and Vietnam, armament was being smuggled through Laos near the homes and villages of the Hmong. Since the US could not officially train out of Laos, they hired the local people, the majority of whom were Hmong, to disrupt the Vietnamese operations in Laos. By 1969, more than one-hundred-and-ten-thousand Hmong had become displaced by the fighting.

When the United States pulled out of the war in 1975, the Hmong were left stranded. By then, North Vietnam had won and become a communist country and Laos was falling in the same way. This meant that the American allies became targets for more oppression and "reeducation," and were eventually killed. To avoid these fates, thousands of Hmong became refugees and sought asylum in Thailand refugee camps across the border. This act included hiding in the forest for months, leaving behind elders who could not make the trek, leaving families who could not leave their villages, and giving opium to babies to keep them from crying for fear of being found by the communist soldiers. Many citizens were killed during this time because there were no more troops and soldiers left to defend them. Almost all crossed the Mekong River which is as wide, deep, and turbulent as the Mississippi. Many families drowned together trying to cross over to Thailand. There were also many communist spies who would betray the refugees and turn them over to the communist soldiers; there are villages of people who were trapped at the shores of Laos and murdered. From there, the US and other ally countries allowed the refugees to immigrate to their respective countries. Only about thirty-thousand Hmong were able to seek and receive asylum in the United States."

Meg: **"That is so tragic! In the information we've read it's said that the Hmong people may have been feeling guilt for leaving their homeland. Can you tell us about the cultural or spiritual beliefs related to family or ancestors?"**

Mai Vang: "I'll speak on this generally because not all Hmong practice only the Hmong spiritual belief. There are diverse religions in the community including Hmong Christians and Catholics and those who have a variation of ancestor and animist worship. I have heard anecdotes from Western doctors and Hmong individuals alike that the Hmong may have felt guilty and therefore it could have affected their health. And that could be true, but I am not sure that guilt can be directly associated with Hmong spiritual beliefs as there are many reasons for those who survived (Hmong and non-Hmong) to feel guilty after war or a traumatic incident. Though, it could have contributed to a lot of the stress related to Hmong beliefs.

For the Hmong, the war experience includes leaving behind loved ones, losing children and loved ones to the war, leaving behind their homes and gardens which they worked on their entire lives, leaving their farm animals behind, and witnessing tragedy and death along their journey. That said, there is a strong belief in, and traditions related to, ancestor worship."

Kelly: **"Do you know of anyone in your community who passed away from the congenital heart problem that Wes Craven cites as the inspiration for *A Nightmare on Elm Street*?"**

Mai Vang: "I don't personally know those who passed away in their sleep; however, our community is small so I know friends and family who know someone personally who did. There was a wave of Hmong men who died in their sleep even though they seemed to be perfectly healthy before. And anecdotally it seemed that it was tied to their grief. This sleep death has been tracked by the Department of Health. This type of sudden death wasn't isolated only for the

Hmong community; it apparently is a phenomenon in Southeast Asian men in the United States and in Southeast Asia."

Meg: **"Have you, personally, ever experienced sleep paralysis? If so, can you describe your experience?"**

Mai Vang: "Yes, I have experienced sleep paralysis. It's a really scary feeling. You are laying there and something is there and you want to scream but nothing comes out. Sometimes it is difficult to breathe. You cannot move your arms and legs, and no matter how much you struggle, your body doesn't move at all. I had a lot of this experience as a child and teen. It was so bad at that time, that I now have a habit of not falling asleep on my back. It seemed that was the sleep position when most of my experiences occurred."

Kelly: **"That is so scary!"**

As Mai Vang mentioned, in Hmong culture, sleep paralysis is prevalent. It is understood within the culture to be caused by a nocturnal pressing spirit, *dab tsog.* Sleep paralysis is a state associated with the inability to move that occurs when an individual is about to fall asleep or is just waking. Those who have experienced sleep paralysis report a feeling of someone in the room with them, pressure on their chest, and an over-whelming fear.

Hundreds of years ago sleep paralysis was thought to be a visit by an evil entity who wished to crush the life out of its victim. How did, or do, cultures all over the world try to treat or prevent sleep paralysis? Greek physicians in history treated sleep paralysis through phlebotomy, or drawing blood, and a change in diet. Chinese people usually

REM sleep usually happens ninety minutes after falling asleep.

approached the condition by employing the help of a spiritualist. Italians, on the other hand, believed sleeping face down and placing a broom by the door with a pile of sand on the bed would help prevent it.[4]

Currently, in medicine, sleep paralysis has been attributed to such conditions as post-traumatic stress disorder, anxiety, depression, and irregular sleeping habits. Experts explain that if we wake too quickly from rapid eye movement (REM) sleep, where there is no motion or muscle activity, the brain keeps us temporarily paralyzed. To prevent episodes of sleep paralysis, it's recommended to get more sleep, avoid drugs and alcohol, and limit caffeine and electronics before bed. It's also important to remember when experiencing sleep paralysis that it is temporary and will pass. Easier said than done, no doubt, but hopefully concentrating on that fact will help the time pass without increased fear. If only following these tips could have prevented Freddy Krueger from wreaking havoc on Elm Street!

Another aspect of *A Nightmare on Elm Street* was based on a real-life experience. When Wes Craven was a child living in Cleveland, Ohio, he heard noises on the sidewalk outside his second-story window. He described, "it was a man in an overcoat and a sort of fedora hat. Somehow, he sensed that someone was watching, and he looked right up and into my eyes." Craven left the window but went back again to look and the man was still there, staring up at him. "The thing that struck me most about that man [in Cleveland] . . . was that he had a lot of malice in his face. He also had this sort of sick sense of humor about how delightful it was to terrify a child."[5] The real-life inspirations for *A Nightmare on Elm Street* definitely prove that sometimes truth can be stranger than fiction—and perhaps even scarier.

SECTION TWO
SERIAL KILLERS

CHAPTER FOUR .

PSYCHO

Year of Release: 1960
Director: Alfred Hitchcock
Writer: Joseph Stefano
Starring: Anthony Perkins, Janet Leigh
Budget: $806,947
Box Office: $50 million

everal years before Alfred Hitchcock's *Psycho* (1960) caused audiences to collectively gasp, a real monster stalked the rural expanse of Wisconsin. He was the worst sort of villain, a man able to portray to his neighbors that he was quiet, sweet, and even a bit slow. It was a shock to those in the desolate farm town of Plainfield when the heinous truth was revealed in the winter of 1957. Ed Gein, trusted to occasionally babysit his fellow farmers' children, had managed to hide a depravity so abnormal that it would inspire the creation of some of film's most notorious monsters. Surprisingly, Gein is the spark that ignited both the timid and proper Norman Bates of *Psycho* as well as the mute and brutal Leatherface of *The Texas Chainsaw Massacre* (1974). It is this duality of the light and the dark that has placed Norman Bates in the upper echelons of horror film fiends.

Robert Bloch's novel, *Psycho*, a novelized account of Ed Gein's house of horrors, was published in 1959. Once Hitchcock read the story of Norman Bates, a man with a mommy complex which charged his sick compulsion to kill women, the famously fastidious director knew he had the subject for his next picture. So certain that his film version would shock the proper filmgoers of the era, Hitchcock ordered his assistant to buy up copies of the novel in order to keep the twists a secret.

The climb for Alfred Hitchcock's production was a steep one. He was alone in his vigor for the macabre project, eventually financing the movie with his personal money. Paramount Studios, reluctant to be a part of *Psycho*, finally agreed to distribute the film if Hitchcock waived his director's fee. The production of *Psycho* itself was rife with Hollywood drama, later becoming the subject matter of the film *Hitchcock* (2012) starring Hannibal Lecter himself, Anthony Hopkins. As depicted in the Hitchcock biopic, as well as the podcast *Inside Psycho* (2017), the making of *Psycho* became a watershed moment in the film industry, particularly in the relationship between auteurs and the stolid protectors of decency, the Motion Picture Association of America (MPAA). From Hitchcock's bold insistence that the MPAA allow what was considered highly violent imagery for the era, to the first filmed flushing toilet in cinematic history, *Psycho* pushed boundaries. Star Janet Leigh (Marion Crane) recounted her time on the set in the 1995 book she co-wrote entitled *Psycho: Behind the Scenes of the Classic Thriller*. Leigh recalled Hitchcock's unconventional methods, including how she once walked into her dressing room to discover the director had secreted the "mother corpse" there to scare her. Whether this was to test the efficacy of the prop, or to keep Leigh in character, she never knew.[1]

The impact of Hitchcock's arguably most famous and well-received film ripples on. *Psycho* boasts several of the most iconic sequences in history, including the notorious shower scene. Marion's shocking murder, only a third of the way through the film, sparked a fear of bathing for generations. No one can watch *Psycho* and then jump into the shower without a creeping dread! This vulnerability, of being murdered in a common place by a man who most would deem safe is why *Psycho* transcended the horror films of its time. It is Norman Bates's (Anthony Perkins) sweet, shy nature that disarms Marion, as well as the audience. He is not a typical monster, a skulking vampire with glistening fangs. Norman appears sincere in his banality, the sort of man who is not sexually appealing to Marion, nor threatening. It is in this aspect that Norman is so similar to his real-life influence, Ed Gein. Both men were able to maintain a facade that ultimately crumbles into brutal violence.

Though one trait of Norman Bates that is dissimilar to Gein is the presence of dual personalities. Gein was indeed consumed by the significance of his dead mother's opinions and rules, but he was never known to actually embody her voice or personality. Norman, on the other hand, seems to slip between two distinct selves. While Marion Crane waits in the motel's parlor, she hears Norman argue with his mother. Not the ravings of a madman, but two voices, two consciousnesses. This scene is particularly chilling when watched with the full knowledge of the twist. And in the end, we once again see Norman's alternative personality modeled after his domineering mother. She takes over Norman's body. Hitchcock even employs a superimposing film technique to illustrate this phenomenon. The audience is then left with Norman's lingering, creepy smile which finishes the film.

This strain of the good and the bad harkens back to the original literary monster of dichotomy, Robert Louis Stevenson's *Strange Case of Dr. Jekyll and Mr. Hyde* (1886). In a 2007 study, Kieran McNally described Stevenson's novel as inspired by several sources: "it drew on theological and literary influences concerning humanity's primitive capacity for good and evil, Charles Darwin's simian conclusions from *The Descent of Man*."[2] It's fascinating to note that evolution was a key point in the contrast of female and male serial killers (see Chapter Three (pg. 13)) and is

Is it possible for people to have more than one personality?

also integral in the understanding of the age-old contrast of good and evil. While the father of evolutionary theory, Charles Darwin, recognized this pull of both light and dark through a scientific lens, Stevenson poetically described it for readers of Victorian literature: "with every day, and from both sides of my intelligence, the moral and the intellectual, I thus drew steadily nearer to the truth, by whose partial discovery I have been doomed to such a dreadful shipwreck: that man is not truly one, but truly two."[3]

Stevenson's novel was later adapted to film in 1931 by Paramount Pictures (the same company reluctant to get involved with Hitchcock's

Psycho). *Dr. Jekyll and Mr. Hyde* earned its star, Fredric March, an Academy Award for his portrayal of a man harboring the vast schism of good and evil. Ten years later, MGM tried its hand at the popular horror story, adding headliners Spencer Tracy and Lana Turner. 1941's *Dr. Jekyll and Mr. Hyde* was directed by *Gone with the Wind's* (1939) famed Victor Fleming. Just as Mr. Hyde might have schemed for nefarious gains, Fleming and the producers of the second film worked to hide copies of the 1931 version in order to lessen their competition. This is oddly reminiscent of Hitchcock buying up Robert Bloch's *Psycho*, leading us to wonder what Darwin would make of these directors' questionable actions in order to come out on top.

Norman Bates and Dr. Jekyll got us thinking about the actual phenomenon of fractured self. As a teenager, I (Meg) became fascinated with this topic, delving into nonfiction books like *Sybil* (1973) by Flora Rheta Schreiber, and *The Three Faces of Eve* (1950) by Corbett H. Thigpen and Hervey M. Cleckley. Both of these books were later turned into biopics, dramatically portraying the anguish of women who were an amalgam of numerous personalities. Fictionalized split personality has long been a popular trope in horror films, from *Identity* (2003), to *Secret Window* (2004), and *Split* (2016). While it has undoubtedly been perpetuated in forms of entertainment, the notion of split personalities, clinically known as dissociative identity disorder (DID), has been one wrought with controversy. Many in the medical community believe it to be a pseudo-condition, while others fight for it to be legitimized in both the medical and cultural spheres. This begs the scientific question, is dissociative identity disorder a proven condition? And what are the attitudes toward it in the psychiatric community? And if it is authentic, would this mean Norman Bates wouldn't be responsible for his crimes? To find out more, we spoke with Mayo Clinic psychiatrist William Leasure, MD:

Meg: **"Can you first tell us a little about your practice and the sort of work you do on a daily basis?"**

Dr. Leasure: "I work in a practice called Integrated Behavioral Health. The name comes from our integration within the primary care practice, where we provide mental health services. The practice is an outpatient practice and we work hand-in-hand with primary care providers. The patients have a variety of mental health problems, with the most common being depression, anxiety, bipolar disorder, ADHD, schizophrenia, substance use disorders, and personality disorders. Patients are typically referred to me, a psychiatrist, for diagnostic questions and treatment with psychotropic medications. Most patients are seen by me on a short-term basis and then returned to their primary care provider for ongoing management."

Kelly: "In the film *Psycho*, Norman Bates notoriously embodies his mother by wearing her clothes and speaking in her voice. He kills 'as' his mother and then is shown to not remember this act. Would this lead you down the road of considering a type of dissociative disorder?"

Dr. Leasure: "The behavior does in some ways suggest dissociation; however, its portrayal depicts a dramatized version of dissociative identity disorder, which is a controversial diagnosis that I am highly skeptical of. More common than embodiment of another's personality are dissociative symptoms which may take the form of depersonalization (experiences of unreality or detachment from one's mind, self, or body), derealization (experiences of unreality or detachment from one's surroundings), or dissociative amnesia. These symptoms are not uncommon in individuals who have experienced trauma. They are typically short-lived and do not involve taking on another's identity, dressing as them, talking, or acting as them."

Meg: "As you mentioned, dissociative identity disorder (formerly multiple personality disorder) is a controversial diagnosis. *Psychology Today* calls DID "unstable, open to debate and hard to pin down." Was it covered in your education? If so, how has it been treated in the psychiatric community?"

Dr. Leasure: "DID is indeed a controversial diagnosis. Most modern mental health providers, including psychiatrists, strongly question the validity of the diagnosis, or outright believe it is a bogus diagnosis that is better accounted for by other problems, which may include substance use, neurologic disorders, PTSD, personality disorders, or malingering. In my training we received essentially no training related to DID diagnosis, other than skepticism of its existence and acknowledgment that diagnostic criteria for the diagnosis exist in the *Diagnostic and Statistical Manual of Mental Disorders* (DSM), which is now in its 5th version. There is a common acceptance that a large number of individuals who are purported to have DID have a history of often severe physical or sexual trauma in childhood."

Kelly: **"Based on your knowledge and experience, where do *you* stand on the controversy of DID?"**

Dr. Leasure: "I am highly skeptical of the diagnosis and close to the line of thinking it is completely bogus. I am hesitant to state outright that it doesn't exist, but if it does, it is exceedingly rare to the point that most clinicians have no practical knowledge or experience with it."

Meg: **"Wow, fascinating! I knew it was rare, but that puts it into perspective. So, imagine Norman Bates is in your office seeking help. He describes an extreme fixation on his mother, as well as moments of amnesia, and let's not forget how this dangerous fixation is now spreading to all the women in Norman's midst (as it did with the character's real-life inspiration, Ed Gein). What sort of modern medicine/therapy could you provide Norman?"**

Dr. Leasure: "My first priority would be assessing safety. I would be interested in hearing if at any time he had thoughts of harming his mother or anyone else, or if they were otherwise at risk. I would be interested in the nature of his fixation with them to help in this determination. Assessing for any abnormalities in his thinking, such as delusional beliefs, would also be important. With the periods of amnesia, I would be interested in determining if he'd had any seizures, head injuries, or other medical problems or medication side effects that might have caused amnesia. I would also be interested in assessing his use of substances to determine if they might have contributed to the amnesia or changes in his thinking. Potential treatment is difficult to speculate about without knowing more about potential symptoms or diagnoses. Certainly, if there were safety concerns due to a mental disorder then psychiatric hospitalization might be warranted. As for medication, if he had psychosis, then I would consider an antipsychotic medication. However, without knowing more information, it's not clear to me how a medication might be effective. Psychotherapy could be a consideration."

Kelly: **"Too bad Norman didn't get help! How does modern medicine/therapy differ from what would have been offered to him in 1960?"**

Dr. Leasure: "I don't know which treatment might have been more likely for Norman Bates at the time. Treatment in 1960 might have gone down a couple of paths. One might have employed psychoanalysis to try to get at the root cause of the problem and remedy it. In Mr. Bates's case there likely would have been a focus on his development, upbringing, early childhood experiences, trauma, relationships, etc. Given the fixation on his mother, his relationship with her would have been explored in detail. Another potential treatment avenue could have been institutionalization where he might have received heavily sedating medication. Medication options were limited and many of those used were heavily sedating. Until the sixties, long-term institutionalization was not uncommon. Electroconvulsive therapy (ECT) was also a treatment option, although I'm not sure if it would have been used in this situation. It fell out of favor in the 1960s before becoming more commonly used starting around the 1980s."

Kelly: **"If Norman did indeed have a sort of dissociative amnesia, would you consider that he would or would not be to blame for the crimes he committed? Could you tell us a little about legal** culpability for people suffering from any sort of similar disorder or affliction as Norman?"

Dr. Leasure: "Forensic psychiatry is an area I have only limited knowledge of. In his case, if he was determined to be in a dissociated state at the time of the crime with little or no volitional control over his actions, then I think not guilty by reason of insanity (NGRI) would be a reasonable outcome of a trial. I can't really speak to the question of legal culpability other than to say it is variable and often up to the jury or a judge to decide. Forensic psychiatrists or psychologists may provide expert testimony, but I believe there are restrictions on how far they can go in making determinations about what should happen as a result of a trial. If someone is found NGRI they are almost always committed to a psychiatric facility for treatment, often with the idea of restoring them to a state in which they are no longer a danger to society. Often they are kept there longer than might clinically be thought necessary because judges are reluctant to release them for fear the patient might commit a crime and the judges receive blame."

Meg: "**And lastly, what do you think of the depiction of mentally unstable, violent people in film? Are there any that stand out to you as chilling in their accuracy? Or upsetting in their generalization?**"

Dr. Leasure: "Mental illness is often portrayed in terrible ways in film that perpetuate stereotypes, such as that people with mental illness are dangerous or crazy. They also sometimes portray purely illegal or psychopathic behavior as due to a mental illness. I would view *Psycho* as potentially guilty of that. Treatments are also sometimes portrayed in a negative light, such as ECT in *One Flew over the Cuckoo's Nest* (1975). A couple of examples that, while perhaps not fully capturing the mental illness, at least highlight the challenges of dealing with it for individuals or families are *A Beautiful Mind* (2001) and *Silver Linings Playbook* (2012)."

It was fascinating to see Norman Bates through the eyes of a profes-
sional. We had heard that DID was a rare and controversial diagnosis,
but hearing Dr. Leasure's thoughts solidified how few people suffer from
this (possibly bogus) condition. If Norman had had the opportunity to
be treated by a psychiatrist like Dr. Leasure, his story may have ended
very differently.

CHAPTER FIVE

THE TEXAS CHAINSAW MASSACRE

Year of Release: 1974	
Director: Tobe Hooper	
Writer: Kim Henkel, Tobe Hooper	
Starring: Marilyn Burns, Gunnar Hansen	
Budget: $80,000	
Box Office: $30.9 million	

Ed Gein died the year I (Meg) was born. As a teenager I happened across a paperback about Gein at a library sale. This book, yellowed and full of black and white, glossy photographs, sparked my interest in true crime. It revealed the monster behind the movies I loved. Ed Gein's story, wholly true and wholly awful, is often described as the igniter of nearly every modern serial killer movie.

For the shaping of *Psycho*, Robert Bloch, Alfred Hitchcock, and Anthony Perkins worked to mold Norman Bates from the proverbial clump of Ed Gein clay. They focused on the duality of Gein's life and his fixation on his dead mother. *Psycho*, while groundbreaking, was a film of its era. In the 1960s, Hitchcock and his film contemporaries simply could not delve any deeper into the truly horrific reality of Ed Gein's crimes, but fourteen years after the iconic film's release, American moviegoers were becoming more conditioned to watch and discuss violence due to the brutality of the Vietnam War being splashed across the nightly news.

Tobe Hooper's *The Texas Chainsaw Massacre* (1974) premiered to the children of those who had watched *Psycho* in the theater, providing this next generation with a cinematic yet starkly authentic look at what had been discovered in that farmhouse in Wisconsin in 1957.

It was on the chilly November evening when Bernice Worden, proprietor of the local hardware store, went missing. Bernice's son, Frank, discovered Worden's Hardware store unattended. A telltale streak of blood on the linoleum spoke of a violent end. In mere hours an unlikely suspect formed in the minds of the Plainfield police. Witnesses had seen local bachelor, Ed Gein, nervously enter and reenter the hardware store more than once that day. Intrigued by this peculiar behavior, Frank, a deputy sheriff, had Gein rounded up by fellow police. Gein had been enjoying supper with another local family when the cops came calling.

What followed, the subsequent investigation of Gein's house, lives on as one of America's most macabre true legends. "Police found the headless, gutted body of [Bernice Worden] at Gein's farmhouse." The unfortunate woman had been treated like a deer. Gein had used his hunting expertise to "dress" her body. "Upon further investigation, authorities discovered a collection of human skulls along with furniture and clothing, including a suit, made from human body parts and skin. Gein told police he had dug up the graves of recently buried women who reminded him of his mother."[1] The search of the house of horrors continued to unnerve the seasoned investigators "yield[ing] more shocking discoveries, including organs in jars and skulls used as soup bowls."[2] While several of these body parts were found to be that of another missing local, Mary Hogan, most had been stolen from the nearby Plainfield Cemetery.

While we watched *The Texas Chainsaw Massacre* with the full knowledge of Ed Gein's influence on the birth of Leatherface, it was easy to see the comparison of that house in Wisconsin with the film's fictional lair. When Sally Hardesty (Marilyn Burns) bears witness to the haunting rooms full of both human and animal parts, it harkens back to how Gein's house must have appeared to those sifting through the horrific finds.

"[Hooper] had heard of Ed Gein, the man in Plainfield, Wisconsin, who was arrested in the late 1950s for killing his neighbor and on whom the movie *Psycho* was based. So when they set out to write this movie, they decided to have a family of killers who had some of the characteristics of Gein: the skin masks, the furniture made from bones, the possibility of cannibalism."[3] Gunnar Hansen, the actor who donned the mask of Leatherface in *The Texas Chainsaw Massacre*, explained that the late Hooper had indeed taken facts from the Gein case to develop the frightening family who kills and tortures the innocent teens.

Despite the fact that several countries banned the movie because of its overt violence, and some American theaters quit showings after complaints, *The Texas Chainsaw Massacre* generated over $30 million domestically. This success led to Tobe Hooper's rise as one of the film industry's most well-known horror directors. He went on to direct hits like *Salem's Lot* (1979) and *Poltergeist* (1982).

Over the decades, the film has been regularly recognized as a bastion in the horror movie canon. Like many horror franchises that came later, (think Jason, Freddy, and Michael Myers) *The Texas Chainsaw Massacre* is less about the screaming teens and more about the monster. "Leatherface seems to be a palpable somebody, a poignantly confused and overwrought monster who can express himself only in a squealing caterwaul."[4] And much like Freddy and his counterparts, Leatherface and his ragtag group of horrifyingly hickish family members have continued to terrorize.

The Texas Chainsaw Massacre is a busy and profitable franchise. There were three sequels from 1986 to 1994, as well as a 2003 remake starring Jessica Biel, another reboot in 2013, as well as two prequels (2006 and 2017). The allure of Leatherface lives on, confronting new generations with the brutal reality of true monsters.

Ed Gein exists not only in the house full of skin and bone in *The Texas Chainsaw Massacre*. Leatherface, like Norman Bates, exhibits traits that can be traced back to Gein, who admitted to wearing his suit of skin, as well as masks he'd fashioned from women's faces. According to an article in *Psychology Today* "Gein had supposedly hoped to transform into a woman, i.e., to become his mother."[5] This act naturally brings Norman Bates dressed in his mother's clothes to mind, but the literal process of wearing another's skin has become synonymous with the fictional monster, Leatherface. Leatherface and Bates also share a similarity in how they treat their victims. For instance, Bernice Worden was found hanging in Gein's shed as though she were the remains of a deer and not a human. This cold indifference of a human, treating them as though they are nothing more than their physical body, is a trademark of Leatherface. He, too, treats his kills as though they are meat. He hangs them from meat hooks and even disturbingly places them in a freezer like we would hamburger patties.

There are a number of ways to classify serial murderers. Holmes and Holmes described several typologies in their work, including hedonistic killers, power-seekers, and more.[6] Other typologies include organized versus disorganized murderers, and process killers versus product killers. Process killers are the serial murderers who enjoy the literal process of taking a life. They often torture and prolong the act of murder. The Golden State Killer, Joseph DeAngelo, apprehended in 2018, is a prime example, as he raped, bound, and strangled his victims. Lust killers like Ted Bundy could also be classified as "process-focused." Ed Gein killed Bernice Worden with a rifle. There is no reason to believe he reveled in the process of murdering her. This "act-focused" type of killing was done in order to gain her body, which for him allowed the fantasy of "transforming into a woman." Fellow Wisconsinite Jeffrey Dahmer, a cannibal and necrophiliac, is another example of a serial killer who did not kill for the act but for the after-effects. Like his cinematic father, Leatherface kills quickly. He does not prolong the act. Although, assuming in order to increase the brutality and suspense, Leatherface was given a chainsaw with which to stalk the dark trails of Texas.

There are a lot of psychological questions to ask about Ed Gein and Leatherface—like why would someone feel compelled to inhabit another's skin? While product serial killers represent the extremely twisted side of this spectrum (collecting human parts!), there are millions of people who practice the science and art of taxidermy. To learn more, we interviewed Lexi Ames, a taxidermy enthusiast, artist, and former apprentice of specimens at Lawrence University:

Meg: **"First, could you tell us about your background? What intrigued you about taxidermy and specimen collection?"**

Lexi Ames: "I have a BA in Biology and Studio Art from Lawrence University in Appleton, Wisconsin, where I focused my studies and art on the history of medical illustration, the portrayal and role of women in science, and the nature of death and decomposition. I enjoy collecting antique taxidermy, bones, medical texts, and odd ephemera.

Collecting oddities and beautiful things has always been a big family to-do in my household, and none of us shy away from the grizzly or grim. I have very clear memories of my father pulling over to show me roadkill to explain that it would turn back to soil, or returning to a deer carcass every few weeks until just the bones remained so that we could bring the skull home

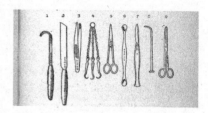

Tools used in taxidermy.

(she's still hanging in my childhood bathroom!). There was never anything malicious in our actions as a family, we were all simply curious of the natural world around us, and comfortable with exploring slightly deeper than most. Of course, I also loved being around living creatures, and had a happy parade of pets, from dogs and guinea pigs to rats and geckos. I loved being able to watch their movements and simultaneously understand what made them move and how they breathed or ate. In a way, I think it helped me develop a healthy appreciation of how beautiful and fragile life is, and a respectful curiosity of death."

Kelly: **"A skull in your bathroom? That's cool! What is the collection process like? Are you in the woods scavenging? Do people let you know about potential pieces to collect?"**

Lexi Ames: "'Scavenger' is probably the most accurate descriptor for what I am! I am constantly on the lookout for a good specimen, and while the woods and riversides are particularly good places, even cityscapes can have excellent bone caches. Window wells tend to trap rodents and rabbits, while nesting birds of prey regularly push bones out of their nests, scattering little treasures all over the ground. I keep plastic baggies and gloves in my car for just such situations.

People love to tell me about dead animals they find. It hit a peak during my bird specimen days at Lawrence. I would patrol the grounds at five each morning to find birds that hit windows the evening before, but I found that the best information came from friends and strangers aware of what I was working on. I'd get a message and sprint out of class mid-lecture to make sure the birds were as fresh as possible, tag them, and bring them straight to my advisor's freezer! My friends and family still love to tell me all about a particular dead thing that they saw, or send a text with a heartwarming "thinking of you" featuring a dead bug or vole their cat dragged into the house. I find it very sweet!"

Meg: **"How thoughtful! And I love that you're prepared at all times! What is a wet specimen?"**

Lexi Ames: "A wet specimen is simply an animal or organ preserved with fluids in a vessel. This can be done using formalin, ethanol, isopropyl alcohol, or another liquid preserve. This process is especially handy for animals with very soft tissues, such as octopus or larval insects, but just about anything can be preserved in a jar! They last very well and it's much simpler than skinning, cleaning, mounting, and detailing a freestanding piece of taxidermy.

Kelly: **"Now I know what all those jars are that we see in so many horror movies! How does the collection of these animals aid in our scientific understanding? Anything surprising you've learned along the way?"**

Lexi Ames: "At one point, all species described by science required a preserved specimen that 'proved' their existence. Collecting samples provides many benefits to science and the public. Collecting city animals hit by cars may, for example, offer valuable information on parasite infestations, food availability, disease concerns, and even minute evolutionary changes! This data can then be used for the benefit of both animal and human residents."

Kelly: **"I just read about this in relation to chronic wasting disease and deer specimens!"**

Lexi Ames: "Museums often have drawer upon drawer of these samples, hidden away from public view. Collecting samples over vast periods of time can also show us how animals are adapting to new challenges like habitat loss or climate change, and help us estimate how future populations may change throughout time. Then, of course, there's the inspiring beauty to these collections. Observers can get very close to the specimens and drink in the detail and structure of the animal in question without worrying about putting the animal into a state of stress. This is especially inspiring to young children, and helps to instate curiosity and wonder of the natural world. Last, it is good to remember that these collections often were taken at a high cost, and that the field of science has greatly expanded its collective opinions on ethicality."

Meg: **"I love how you describe the balance of science and art! What is one piece that you were super excited about? A find that was a big one for you."**

Lexi Ames: "One spring, I was hiking with a friend in rural Wisconsin when we came across an entire skeleton of a massive ten-point buck. It's highly unusual to find such a treasure; scavengers hadn't removed any pieces from the site, and aside from the mat of hair he laid on, the bones were nearly clean. He was laid out in a shady, fern-filled valley, with only his antlers visible above the foliage. We gathered up every last bit except for the hooves as they had become too soft, and carried them several miles back to the car in our sweatshirts and raincoats. We call him Heathcliff, and he hangs in my childhood home next to Cathy, the doe skull my father and I collected when I was four years old."

Kelly: "Heathcliff and Cathy, love the *Wuthering Heights* reference! What sort of reactions do you get about being a taxidermy enthusiast and collector? And what are some inaccuracies people believe that are not true about taxidermy and bone/fur collection?"

Lexi Ames: "While most people find it engaging and enjoy asking questions about the animals and process, some people are very disturbed by it, either in reaction to the welfare of animals, or the fear of death and the nature of dead bodies. As an animal lover myself, I understand where the concerns of those in the former category come from. Hunting practices and meat consumption can be highly divisive topics, and I try to keep up with facts on both sides of the equation to facilitate conversation and keep my own mind open. The animals and bones in my collection have come to me in mostly gentle ways, usually after a natural death. I believe people in the latter category, those afraid of the dead body itself, reflect the state of America's disengagement with death as a society. We interact with death very little, and sterilize the experience when we do. However, new death movements here in the states, such as greener burial practices, are making people slowly more comfortable with mortality. Even pet taxidermy is becoming more common. I think we are seeing a turning point in accepting reminders of death, like taxidermy and open conversations on the topic, back into our homes. Maybe it can even unite the country! Game taxidermists and pet taxidermists coming together could be a beautiful thing.

Another common misconception is that it is a highly dirty and unfeeling practice and composed mostly of men. While there are precautions to be taken, and animal bodies do harbor bacteria and foul smells, the process requires a steady, gentle hand. A deep respect for the animal's remains is required to make the creature life-like again. And the field is far from lacking women—almost all the taxidermists I follow on social media identify as women! It's a very exciting time for the art form."

Meg: "**You managed a bird and small mammal specimen collection in college, do you think that experience impacted your views on death?**"

Lexi Ames: "A quick background on the collection: The Lawrence University ornithology collection is made up of some three-hundred birds from over two-hundred years of collection (we also had many small mammals like voles and bats). I digitized the collection, and obtained the proper paperwork from the Wisconsin DNR to collect birds killed in window-strikes to add to the assemblage. I was also trained to preserve their bodies as study skins. A study skin is the bird's intact pelt, wings, head, and legs with the internal organs, fat, and muscles removed. It looks like a bird sleeping neatly on its back with the wings tucked under and the eyes stuffed with cotton. My views of both death and conservation were sent into overdrive while working with the birds. The sheer numbers being killed on a campus the size of just a few city blocks was astounding. Sometimes as many as four birds would be killed all at once when their flock migrated through and became confused by the glass, and our freezer was constantly full. The diversity of species was also quite impressive, and made me want to spread word about doing more to protect the birds as they passed through. The preservation process of turning a body into a study skin is also quite intimate, and a very hands-on process. Stated simply, the bird's body and soft tissue must be turned out of the skin, leaving behind a sort of bird jacket. The inside of the "jacket" is then cleaned, dried with sawdust, stuffed with cotton, stitched up, and left to dry on its back. It's a delicate procedure, and requires speed and gentleness. I always felt very lucky to be able to witness the inner workings of these little creatures. They're highly delicate, and each was an individual, like a person. Some would have particularly thick wing muscles or bright feathers compared to other members of their species, or a weaker, less healthy bird may have evidence of mites and ticks. I was always moved by the experience."

Kelly: "Fascinating! Meg is not okay right now because she has a fear of birds."

Meg: "I do!"

Kelly: "Do you think horror films (Norman Bates was an avid taxidermist) or the media in general have given people a negative view of taxidermy? Are there any films that have shown it in a positive light?"

Lexi Ames: "Oh my goodness, yes. As a big horror film fan and true-crime junky myself, I can't help but be riveted by a story about someone like Ed Gein or Jeffrey Dahmer that involves collecting bones in early life. Obviously, there is something greatly different in the motivations of serial killers versus that of your local taxidermist or anatomy enthusiast, but what is it, exactly? I think it becomes very difficult to see the difference and draw the line of distinction if you are someone who isn't compelled by anatomy. If someone hasn't grown up going to natural history museums frequently, their only lens of taxidermy may be through the horror genre. For me, interacting with anatomy is like interacting with a great art form, and I think we are typically trained as a society to see it as the exact opposite of that.

I really racked my brain to think of 'nice' taxidermists in film or television, and had a difficult time drumming up any! The closest I could come was Vanessa Ives in the series *Penny Dreadful* (2014–2016). Of course, every character in this show is meant to be inherently creepy, but Vanessa is a particularly strong and curious person, and even with her dark sides she is impossible not to admire. In a rare scene that depicts Vanessa as happy and contented, she is shown dabbling with taxidermy as a young child. Although her prodigiousness in the craft is completely unbelievable (she's like a ten-year-old nineteenth-century Martha Stewart of death!), it's charming to see her and her young friends explore the natural world through the art before everything in their lives is swallowed by betrayal and darkness. Overall though, it's clearly entertaining to be creeped out by a character who curates the dead. I can't imagine that trope disappearing, especially as there are too many real-world examples in public memory, but perhaps it can expand into normalcy in other genres. It would be great to see taxidermists and death workers portrayed in entertainment that are just plain boring and unremarkable."

Kelly: **"Or without the stereotype of a morgue worker eating over the dead bodies!"**

Thanks to Lexi Ames, we learned so much about taxidermy, specimen collection, and the intersection of science and art which feels apropos for a book about the science in film. We don't think we'll ever look at roadkill the same after our enlightening interview!

CHAPTER SIX

THE SILENCE OF THE LAMBS

Year of Release: 1991	
Director: Jonathan Demme	
Writer: Ted Tally	
Starring: Jodie Foster, Anthony Hopkins	
Budget: $19 million	
Box Office: $272.7 million	

There are two monsters in the acclaimed, Academy Award–winning movie *The Silence of the Lambs* (1991). If we were to use the serial killer classifications laid out by Holmes and Holmes on these fabricated killers, Hannibal Lecter (Anthony Hopkins) and Buffalo Bill (Ted Levine) would fall into similar categories. Both men are compelled to kill for a product. Lecter eats the bodies of his victims, and as he's a class act, he serves them up with the famous "fava beans and a nice chianti!" While Buffalo Bill, far less sophisticated, kills women of a larger girth in order to inhabit their skin. This, of course, evokes the anti-hero of this section, Ed Gein, who shared in precisely the same practice. Gein purposely chose larger women. Both Mary Hogan and Bernice Worden were known to be plus-size, as are Buffalo Bill's markedly younger victims. Surrounded by the bones, organs, and skin of his victims and those he grave robbed, Ed Gein engaged in any number of horrific acts. It's easy to let our macabre imaginations run wild. Yet, there is one misnomer that has been perpetuated since his capture: there is no proof that Ed Gein was ever a cannibal. Because of their fixation on female skin, Buffalo Bill and Ed Gein are inexorably linked. Any link between Gein and the well-educated and poised Hannibal Lecter would be a tenuous one. In fact, it was another true monster who resulted in the creation of the notorious Dr. Lecter.

Before he was a famous author, Thomas Harris was a journalist for *Argosy* magazine. In his early twenties, he was tasked with interviewing Dykes Askew Simmons, a murderer on death row in Mexico. During Harris's visit to Nuevo León State Prison, he also spoke with the prison's doctor regarding an injury Simmons had suffered while incarcerated. This exchange with the physician struck Harris as intriguing, especially when the doctor brought up rather philosophical musings about "the nature of torment."[1] When Harris finished up his interview, he asked the warden how long the doctor had been employed by the prison. The Warden informed the young journalist that the doctor was a murderer, sharing the same fate as his patient, as he was also on death row. This man, who Harris has not named publicly, but is widely believed to be Dr. Alfredo Balli Trevino, had been convicted of the murder of his lover, Jesús Castillo Rangel. According to *The Sun*, in October of 1959 Balli Trevino slit his boyfriend "Rangel's throat with a scalpel in a crime of passion before finishing him off in the bathroom. Balli Trevino cut the body into pieces and buried them to hide the grim crime."[2] This poised man of science, who Harris would later describe as having "a certain elegance about him" had been the perpetrator of a brutal crime. It was this dichotomy, not unlike the good versus bad conundrum of Norman Bates, which Harris drew from when writing *Red Dragon* (1981). Balli Trevino served twenty years before his sentence was commuted in 1981. He spent the rest of his life, until his death in 2009, serving the poor in Monterrey, Mexico. Ironically, the man who inspired Harris to create Hannibal Lecter was known in his community for his large heart, and for not charging the sick and elderly for his medical expertise.

Harris's novel *Red Dragon*, containing the first iteration of Dr. Lecter, was a successful novel that spawned a less successful film; *Manhunter* (1986). While *Manhunter* underperformed at the box office, this did not dissuade Orion Pictures from capitalizing on the wildly popular novel sequel *The Silence of the Lambs* (1988). Orion's gamble paid off, the film released on Valentine's Day 1991 to both critical acclaim and a solid box office showing.

While Dr. Lecter and Buffalo Bill represent the darker side or the "Mr. Hyde" of *The Silence of the Lambs*, Clarice Starling (Jodie Foster) is the film's reasonable, empathetic "Dr. Jekyll." Foster's performance as a newbie FBI agent on her first, crucial assignment rivals that of Hopkins's. She won an Academy Award for her efforts, and to this day Starling stands out as one of the more well-liked and memorable horror movie protagonists. She is bold enough to challenge Lecter with his greatest weapon, his intelligence, when she says "you see a lot, Doctor. But are you strong enough to point that high-powered perception at yourself? What about it? Why don't you—why don't you look at yourself and write down what you see? Or maybe you're afraid to." It is her back and forth with Lecter, their oddly dynamic dance of wits, that catapults this movie into so many "best of" lists, horror or otherwise. The American Film Institute has consistently showered *The Silence of the Lambs* with accolades, including #65 in the best movies of all time, and Hannibal Lecter nabbed the top spot as the #1 villain in AFI's 100 Years... 100 Heroes and Villains.

What is it about Hannibal Lecter that resonates with audiences? And particularly as embodied by Anthony Hopkins? As Brian Cox's Hannibal Lecter in *Manhunter* didn't seem to affect a similar fervor. In his study "The Devil Made Me Do It: The Criminological Theories of Hannibal Lecter" J. C. Oleson posits, "there are a number of plausible explanations for Lecter's uncanny popularity. It has previously been suggested that the character of Hannibal Lecter may fascinate the public because he is enigmatic, fitting several models of serial homicide, while defying others. The allure of the character may also be linked to Hannibal Lecter's status as a criminal genius."[3]

Is it Hannibal Lecter's high IQ that captivates us? Hopkins's portrayal as a man with a dense vocabulary certainly subverts our expectations of the mute thugs of our nightmares. Jason Voorhees and Michael Myers were not known for their sparkling dinner repartee. Perhaps it is this stark difference that absorbs our attention. In fact, a serial killer with an abnormally high IQ is not just a fiction of Thomas Harris's. Ed Kemper, a real serial murderer popularized by the horrifically disturbing performance by Cameron Britton in 2017's *Mindhunter* (not to be confused with *Manhunter*!) tested on the IQ scale at 145.[4] This would mean he'd passed "highly gifted" and was on the low end of "genius."[5] This "genius" used his cunning for evil, killing strangers and family members alike. Rodney Alcala, known to the world as "The Dating Game Killer," responsible for numerous murders of young women across the country, is estimated to have an IQ of about 170. This would mean Alcala is of the "highest genius" and may explain how he was able to escape capture by the police, change his identity, and study at NYU as a wanted man. Another "gifted" serial killer is the infamous cannibal Jeffrey Dahmer, who hovered on the IQ scale near Kemper at 144. While Lecter and Dahmer obviously share in their fixation on eating human flesh, one would have to assume Lecter's IQ would be of the more Alcala variety, if not higher.

What is interesting to note, is that all of these murderers are well-known in the public sphere, which lends to the notion that as a society we are fascinated with high-functioning serial murderers. As Oleson maintains, "the public exhibits a seemingly insatiable appetite for true crime, and has exalted many serial killers into its pantheon of infamy, but one serial killer commands the popular imagination unlike any other: Dr. Hannibal 'The Cannibal' Lecter." While his superior brain enhances Dr. Lecter's charm, there are other facets to his villainy. In his article for *Psychology Today,* Dr. Scott Bonn attempts to explain Lecter's appeal:

> Like many Hollywood monsters and boogeymen, Dr. Hannibal Lecter is exciting and magnetic because he is completely goal oriented, devoid of conscience and almost unstoppable. Hannibal Lecter is uniquely different than any other Hollywood movie monster or killer, however. Unlike cartoonish characters such as Godzilla or Freddy Krueger, Dr. Lecter is human. He is also brilliant, witty and even charming.[6]

Again, we are faced with the duality of monsters. Hannibal Lecter is both monster and man, an amalgam of the two. This schism has burst forth as a defining component of a memorable movie villain. Bonn continues, asserting that it is this humanity that ultimately generates our mesmer-ization by Hannibal "The Cannibal" Lecter:

> My research suggests that Dr. Lecter's enduring popular appeal and the terror he invokes are due to the fact that he is depicted as a mortal man. In many ways, he is like the rest of us. He bleeds and he feels pain. His humanness makes him a much more relatable and identifiable villain to the public.

As we continue our pursuit of further understanding film's most notorious monsters, we are curious to note if this duality exists in witches, slashers, creatures, and beyond.

With all serial killers, both real and fictional, there are a multitude of psychological questions we could pose, though one word or taboo concept is synonymous with Dr. Lecter. Every viewer of *The Silence of the Lambs* surely must have squirmed in their seat at the mention of Lecter's canni-balistic proclivities. While homicidal cannibals like Hannibal Lecter are rare, there are humans who have found themselves in harrowing conditions which have led to eating their fellow humans. Aside from anomalous serial killers, what is the reasoning behind modern-day cannibalism? According to researchers, "it appears that human cannibalism has been carried out during most prehistoric and historic periods. Several Homo species, including Homo sapiens and other ancestral hominin species, practiced this type of consumption, which is associated with a wide range of behaviors."[7] Cannibalism has existed in every era of humanity, yet it is the modern era, when cannibalism is no longer needed for ritual or nutrition, that strikes the most interest in filmmakers and consumers of media.

The Donner Party is widely known as a prime example of "normal" or non-deviant humans resorting to cannibalism to survive. In the winter of 1846, the Donners, along with eighty members of their settler party, became inexorably trapped in the Sierra Nevada Mountains of California. In what was considered one of the worst winters on record, thirty-nine people perished from the cruel effects of the low temperatures and drifts of relentless snow.

After eating all their provisions and animals, the Donners and their counterparts faced certain death. The great irony of this blight on American history, is that the people struggling to survive in the Sierra Nevada Mountains were dying from hypothermia, *not* starvation.

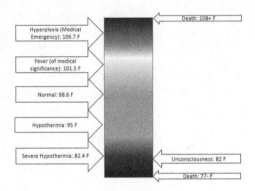

How the body reacts to change
in temperature.

According to Daniel James Brown, author of *The Indifferent Stars Above: The Harrowing Saga of a Donner Party Bride* (2009), because there was little knowledge of how hypothermia worked, those trapped in the mountains believed that starvation was the cause of their discomfort. While they were undoubtedly hungry, they were in reality weeks away from starving to death. But their collective weakening bodies led them to believe they had to eat to live. While this was true later in the winter, it was not the case early on.

Because of this misunderstanding, it was only days into their journey for help that one band of the settlers chose to eat the dead. In order to lessen the psychological trauma of eating their loved ones, they separated into two groups, assuring that they would eat their friends rather than their family. This natural aversion to eating their own mothers, siblings, and spouses, is more in line with how we understand cannibalism as a society. While Hannibal Lecter would relish the eating of a corpse, those poor humans in the Sierra Nevada Mountains ate flesh only when they thought it absolutely necessary. Unfortunately, eating the bodies did little to help, as it was the brutal cold that killed most of the thirty-nine.

More recently, in 1972, Uruguayan Air Force Flight 571 crashed in a remote section of the Andes Mountains. Twenty-eight survivors, many of them composed of a rugby team, were trapped in a similarly hopeless situation as the Donner Party. This incident was chronicled in the 1993 film *Alive*. After an avalanche, ice-cold temperatures, and lack of food, only sixteen survived the seventy-two days of frozen torture. They attributed their survival to making the choice to eat those who perished. Survivor Roberto Canessa explained:

Our common goal was to survive. But what we lacked was food. We had long since run out of the meagre pickings we'd found on the plane, and there was no vegetation or animal life to be found. After just a few days we were feeling the sensation of our own bodies consuming themselves just to remain alive. Before long we would become too weak to recover from starvation. We knew the answer, but it was too terrible to contemplate. The bodies of our friends and team-mates, preserved outside in the snow and ice, contained vital, life-giving protein that could help us survive. But could we do it? For a long time, we agonized. I went out in the snow and prayed to God for guidance. Without His consent, I felt I would be violating the memory of my friends; that I would be stealing their souls. We wondered whether we were going mad even to contemplate such a thing. Had we turned into brute savages? Or was this the only sane thing to do? Truly, we were pushing the limits of our fear.[8]

While this shocking reality caused a clamor of judgment when the survivors told their story to the media, once Canessa and the others explained that those who were dying gave permission for their bodies to be used for nutrition, the families of those who died and were ultimately eaten worked to understand the dire circumstances and forgive those who had survived.

This modern example of cannibalism to survive once again demonstrates our inculcated beliefs that cannibalism is one of the worst atrocities to perform on another. Despite the extreme life-and-death circumstances, there was still hesitation from the survivors, as well as judgment by those who were not there.

In an interview with Bill Schutt, author of *Cannibalism: A Perfectly Natural History* (2017), Beckett Mufson of *VICE* asked why we have an "innate" repulsion of human flesh:

I'm not so sure it's innate. It's deeply ingrained in Western culture. We've been reading this memo since the time of the ancient Greeks. From Homer and Herodotus through the Romans and then Shakespeare and Daniel Dafoe and Sigmund Freud, the snowball kept growing. You're talking over two thousand years. Cannibalism, to these writers, was the worst taboo. Add that to Christianity and Judaism where it's important to keep the body intact and you get the knee-jerk reaction to the very mention of the word we have right now. It has historically been convenient for Westerners to stigmatize cannibalism. If you're Columbus and you can accuse people of being cannibals, then you can treat them like vermin. They're not human to you. You can destroy these cultures. But there are other cultures where they'd be just as mortified to learn we bury our dead as we would be to learn that they eat their loved ones.[9]

It seems that the profound taboo of cannibalism has been perpetuated by Western culture, ingrained in both our religion and our fear of other cultures. Perhaps this is why real cannibalistic killers, in the same vein as the fictional Hannibal Lecter, have garnered such morbid fascination. Although survival is the reason for cannibalism in both the Andes and Donner events, the notion that a human would *choose* to taste flesh is, well, a tough idea to swallow.

SECTION 3
VAMPIRES

CHAPTER SEVEN

DRACULA

Year of Release: 1931	
Director: Tod Browning	
Writer: Garrett Fort	
Starring: Bela Lugosi, Helen Chandler	
Budget: $355,000	
Box Office: $4.2 million	

In 1885, an Irishman named Bram Stoker read a rather enlightening article in the literary magazine, *The Nineteenth Century*. The author, Emily Gerard, wrote of her research on the beliefs of Romanians, entitling the article "Transylvanian Superstitions." She spoke of a creature not widely known in the United Kingdom:

> There are two sorts of vampires—living and dead. The living vampire is in general the illegitimate offspring of two illegitimate persons, but even a flawless pedigree will not ensure anyone against the intrusion of a vampire into his family vault, since every person killed by a nosferatu becomes likewise a vampire after death, and will continue to suck the blood of other innocent people till the spirit has been exorcised, either by opening the grave of the person suspected and driving a stake through the corpse, or firing a pistol shot into the coffin.[1]

Instantly intrigued, Stoker further researched the history and fables of Romania. In his studies he came upon the legends of the notorious, and real, Vlad the Impaler. Also known as Vlad Dracula, this Romanian prince of the fifteenth century was known to be extremely cruel in times of both war and peace. For example, when a group of Ottoman envoys visited Vlad, they made the deadly mistake of not removing their turbans in deference to the prince. "Commending them on their religious devotion, Vlad ensured that their turbans would forever remain on their heads by reportedly having the head coverings nailed to their skulls."[2]

It makes a macabre sort of sense that Bram Stoker combined the vampire folklore he'd learned from Gerard's article with the imposing, historical figure of Vlad the Impaler. This amalgamation of fact and fiction became one of the most recognizable monsters in literary and film history. In 1897, Count Dracula, borne of these darkened legends, first appeared in Bram Stoker's novel, *Dracula*. Over a century later, the mysterious count with a penchant for blood has appeared in hundreds of adaptations.

Although it was not the first film to be inspired by Stoker's novel, the 1931 Universal Studios *Dracula*, starring Bela Lugosi, was the first official film adaptation. The script was taken from the successful Broadway version of the novel, and Lugosi himself was plucked from the stage production to reprise his role as the bloodthirsty Count Dracula. It is Lugosi's portrayal of the Count which has endured: from the slight foreign accent to the air of sophistication.

The early 1930s was a tumultuous time in Hollywood. Silent era films were going out of fashion, yet the industry didn't know how to keep up with this creative change. *Dracula* straddled these two worlds, retaining markers of silent films like intertitles and highly dramatic, theater-like performances. Despite this unevenness, the Universal film was a success. And, soon, Count Dracula became an archetypal monster not only for the readers of Stoker's novel, but for moviegoers all over the world.

Meg and I have both delighted in vampires in horror films. Meg loves *Fright Night* (1985) and I (Kelly) am a fan of *What We Do in the Shadows* (2015). Without both Stoker's novel and Lugosi's 1931 portrayal of Count Dracula, these later films wouldn't exist. As we've re-watched our favorite vampires, we've been wondering about the often-used tropes of this genre. One example is that vampires will be burned to death by sunlight. Oddly enough, this was not depicted in Stoker's novel, but the aspect of vampires not having a reflection did:

Having answered the Count's salutation, I turned to the glass again to see how I had been mistaken. This time there could be no error, for the man was close to me, and I could see him over my shoulder. But there was no reflection of him in the mirror! The whole room behind me was displayed, but there was no sign of a man in it, except myself.[3]

While many of these tropes came from the mind of Stoker, as well as the centuries-old legends of Romania, there is a real medical condition that mimics vampirism, and perhaps informs some of these vampiric idiosyncrasies. Is it possible that medical afflictions inspired the authors and filmmakers who have brought vampires to life? Or even the Eastern European legends which Stoker studied?

One theory of the origin of the vampire is the rare disease called porphyria. Known as "vampire disease," porphyria causes irregularities in the production of heme, a chemical in blood. Some forms of this condition can lead to deposition of toxins in the skin. Sufferers are often sensitive to light as it activates these toxins. While they don't burst into flames like Count Dracula, those afflicted can suffer from disfigurement, including lip and gum erosion. These factors could have led to the corpse-like, fanged appearance that we associate with vampires and their dislike of sunlight. "Porphyrias" also have an intolerance to foods that have a high sulphur content such as garlic.[4] This could have led to the popular myth that vampires are repelled by the stinky vegetable.

One famous case of porphyria was King George III who had acute intermittent porphyria. As portrayed in the film *The Madness of King George* (1994), the most notable symptoms of this type are neurological attacks, such as trances, seizures, and hallucinations, which often persist over days or even weeks. Other famous people who were said to have porphyria include Vlad III, fueling the rumors of him being a vampire, and Vincent Van Gogh.

Another rare type of porphyria, congenital erythropoietic porphyria (CEP), can cause appalling mutilations from the light-activated porphyrins, including loss of facial features and fingers, scarring of the cornea, and blindness. The condition may have been less rare in the past, especially in Transylvania, perhaps giving rise to tales of vampires. While porphyria is a genetic condition, in some cases it can also be caused by environmental contaminants. The most famous episode happened in Turkey in the 1950s, when four thousand people developed a form of porphyria after eating wheat seeds that had been sprayed with a fungicide. Hundreds died, and use of the fungicide was later banned.

This leads us to ask, can porphyria be treated? Studies have shown that spleen removal and bone marrow transplants can be effective. Modern medicine has also found that blood transfusions can help. Interesting to note, the heme pigment is robust enough to survive digestion, and is

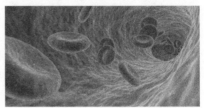

Heme is strong enough to survive digestion.

absorbed from the intestine. This means that, technically, it is possible to relieve the symptoms of porphyria by drinking blood! Who knew Count Dracula was so advanced in medical science?

Another medical explanation for vampires is tuberculosis (TB). This lung disease has been around for centuries and throughout history was spread easily among those living in close, unsanitary conditions. Victims of TB turn very pale, often avoid sunlight, and cough up blood due to the breakdown of lung tissue. Seeing a sick person with blood on their lips and in their mouth was often misinterpreted as them drinking blood rather than coughing it up. TB spreads rapidly and easily from person to person and may have led to the belief that vampires rise from the dead. As multiple family members succumbed to the infectious disease, some believed the undead were visiting at night to drink blood and create more vampires. Nineteenth century New England was gripped by its own "vampire panic" resulting in members of the community exhuming bodies and burning internal organs in order to stop the "vampire" outbreak.

Many famous people throughout history either suffered from TB or knew someone who did, including Edgar Allan Poe, Charlotte Brontë, and Anton Chekhov. Because so many artists had TB it became known as "the romantic disease" and was thought to help its sufferer see life more clearly. TB is still an active disease but much less common due to vaccines and treatment options. Currently, less than two hundred thousand people a year are infected with TB in the US. In the early 1800s it was the leading cause of death afflicting more than 25 percent of the population.

Another theory for the vampire myth is the condition called catalepsy. This disease of the central nervous system leads to a slowing of the heart and breathing rate, with sufferers often appearing to be dead. This condition is portrayed in popular fiction including Edgar Allan Poe's *The Premature Burial* (1844) and *The Fall of the House of Usher* (1839). When a person is afflicted with catalepsy, they are unresponsive to stimuli, may have rigidness, and pale skin. The subject is paralyzed and has no vital signs, which, to most people, would seem like they were dead. This disorder usually lasts a few minutes or hours, but can last up to days in the most extreme cases.

Although Bram Stoker may have started the phenomenon of vampires in literature, the creatures live on in media today. From popular series *True Blood* (2008–2014) to *Twilight* (2008) people continue to be fascinated with the mysterious blood suckers well into the twenty-first century.

CHAPTER EIGHT

NOSFERATU

Year of Release: 1922	
Director: F. W. Murnau	
Writer: Henrik Galeen	
Starring: Max Schreck, Greta Schroeder	
Budget: Unknown	
Box Office: Unknown	

While the 1931 version of *Dracula* was made with the blessing of Bram Stoker's heirs, the 1922 silent German film *Nosferatu* was a blatant rip-off. So obvious, in fact, that Stoker's family and estate sued Prana Film over the copyright infringement. They won, leading to the destruction of almost every copy of *Nosferatu*. The vital word is *almost*, as *Nosferatu*, like its inspiration Count Dracula, is seemingly immortal. Thankfully, the F. W. Murnau picture survived and has gone on to entertain audiences for nearly a century.

Count Orlok, with his pale eyes and spindly fingers, is yet another take on Dracula. The story of *Nosferatu* is strikingly similar to Stoker's novel, complete with its own versions of Jonathan and Mina Harker. And while the film is an admitted copycat, it deviates from the novel in that Count Orlok kills his victims, giving them no chance at eternal life. *Nosferatu* is even credited with first depicting the notion of vampires succumbing to sunlight, which has been inculcated in our modern day understanding of vampires.

F. W. Murnau is, of course, not the first creative to take perhaps a little too much inspiration from another. It is true that *Nosferatu* mirrors itself after Stoker's novel, yet vampires were not a new concept. Stoker stitched together *Dracula* with the before-mentioned Vlad the Impaler. Like all monsters, vampires have a touch of realism, a humanity that anchors them to the natural world. Nearly every movie monster has roots in reality. The consumption of blood has been dated back to ancient civilizations. From Asia, Africa, and beyond there are examples of blood rituals long before Stoker, or Romania, came to be. In Christianity there is even the symbolic drinking of Jesus Christ's blood.

Perhaps inspired by Count Dracula, disturbed people, not unlike *Nosferatu*'s peculiar character Renfield, believe they must have blood to quench their unnatural compulsions. Richard Chase, nicknamed "the vampire of San Francisco" killed six victims in the late 1970s with the purpose of drinking his victim's blood. Before his reign of terror, Chase was hospitalized in a mental asylum where the nurses called him "Dracula" because he had been found "injecting rabbit blood into his veins."[1] Obviously Richard Chase held no supernatural powers, nor a biological reason to drink human blood. Yet, "vampire killers" have been documented all across the world from Japan (Tsutomu Miyazaki) to Germany (Fritz Haarmann), so if humans have found reason, however depraved, to drink each other's blood, we wondered if these instances occur in nature. Are there animals with "bloodsucking" capabilities? And if vampires operate at night, what are the advantages of nocturnalism?

Nocturnal animals are characterized by being active during the night and sleeping during the day. There are numerous animals in nature that are naturally nocturnal including bats, raccoons, and other woodland creatures. Many nocturnal animals have developed or evolved improved eyesight, hearing, and sense of smell. Other benefits of being nocturnal include avoiding the heat of the day, avoiding predators, and avoiding competing for resources. Although the percentage of nocturnal animals is small, scientists have discovered that human disturbance is pushing more mammals to be active at night. According to Kate Jones of University College London, mammals only became active during the daytime after dinosaurs vanished so nocturnalism may be a more "natural" state to be in.[2]

There are a variety of animals in nature whose food source is blood. Vampire bats use their sharp teeth to make an incision on their prey, then lick the flowing blood from the wound. Although they usually feed on livestock and other animals, there have been cases of vampire bats attacking humans. In 2005, bites from rabid vampire bats were blamed for twenty-three deaths in Northern Brazil.[3] Another flying creature, the oxpecker, is an African bird that feeds on the blood left by the bites of insects on a host's hide.

Insects that drink blood are prevalent throughout the world. With over two thousand species worldwide, fleas are the most prevalent parasite found on fur-bearing animals. They actually helped spread the bubonic plague in the 1300s that caused the deaths of an estimated seventy-five to two hundred million people. Ticks are another bloodsucker responsible for spreading

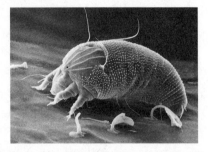

Mites and ticks are nature's blood suckers.

chronic wasting disease and other afflictions. Mosquitos, which are all too familiar to us here in Minnesota, can transmit a number of serious diseases, including yellow fever, malaria, and dengue. This is why they are considered one of the world's deadliest animals. Bedbugs bite with an anesthetizing agent, and while they don't spread disease, they can be difficult to get rid of.

Even some plants consume blood. Carnivorous plants, as shown in *Nosferatu,* are plants that derive at least some of their nutrients from animals or insects. There are at least 583 types of plants that attract, trap, and kill their prey. How can plants attract and trap animals? Some do it with sticky mucus or bacteria pools. Others can pull prey in with quick leaf movements or vacuum-like bladders. *Gross!* The Venus flytrap is probably the most recognizable carnivorous plant and consumes its prey over a one-to two-week period. The plant in *Little Shop of Horrors* (1960), named Audrey II, is similar and has a taste for blood. Can carnivorous plants become large enough to consume humans? Although there are legends of man-eating plants, no claims have been substantiated. In 1874 an account was published in *New York World* of a woman being sacrificed to a tree in Madagascar. The writer's account claimed "... the tendrils one after another, like great green serpents, with brutal energy and infernal rapidity, rose, retracted themselves, and wrapped her about in fold after fold, ever tightening with cruel swiftness and savage tenacity of anacondas fastening upon their prey."[4] In 1891 a story was published in *Review of Reviews* that told of the "vampire vine" in Nicaragua whose desire is "to drain the blood of any living thing which comes within its death-dealing touch." This story, too, was proven to be fabricated.

We concluded that there are indeed many animals with a taste for blood, but always for sustenance. Science and fiction deviate in the vampire's desire for blood. This is where the true horror lies.

CHAPTER NINE

JENNIFER'S BODY

Year of Release: 2009	
Director: Karyn Kusama	
Writer: Diablo Cody	
Starring: Megan Fox, Amanda Seyfried	
Budget: $16 million	
Box Office: $31.6 million	

It is inevitable that fangs and blood will appear when one sees a vampire in their mind's eye. Thanks to Bram Stoker and the thousands of iterations of Count Dracula that have popped up in film, TV, and even on breakfast cereal, nearly everyone in the world shares a similar concept of a classic vampire. But vampirism is not all cloaks, bats, and strained accents. The act of consuming flesh, blood, or even another's spirit or essence in order to thrive has traveled from ancient folklore to the big screen. One such parasitic monster is the succubus. We think of nightmares as disturbing dreams, but the etymology of the word can help us understand phenomena like sleep paralysis and even succubi. "Mare" stems from mara, the Anglo-Saxon word for "crusher," a reference to the sensation of someone or something on the sleeper's chest. The maras of folklore were small imps or goblins, much like the creature squatting in Henry Fuseli's iconic 1781 painting "The Nightmare." Some night visitors may take the form of a male incubus, "that which lies upon" or a female succubus, "that which lies beneath." Each of these demons engage sexually with their victim and have had a presence in Christian demonology throughout history.[1]

In *Jennifer's Body* (2009), a horror-comedy directed by Karyn Kusama and written by Academy Award winner Diablo Cody, a succubus preys on a Midwestern town. High schooler Jennifer Check (Megan Fox) embodies all that the myth suggests. She is beautiful, has an innate sexuality, and most significantly, she must feed on humans in order to survive and retain her beauty. When Jennifer recognizes her body weakening, much like a vampire, she must hunt and kill in order to restore herself. As in the legends of succubi, Jennifer uses her sexual charm to hypnotically lead unsuspecting men to their doom. Because she is much like Count Dracula, who can appear normal and even comforting, Jennifer is herself a snare trap and a monster. Though when she is angered, weak, or hungry, the supernatural awakens within and the true, ugly self rears its head. In *Jennifer's Body*, the hero, Needy Lesnicky (Amanda Seyfried) researches succubi and finds there is a way to kill them. Again, just like vampires, there are rules under which succubi must operate. While they have powers, they also have weaknesses. Every monster has its kryptonite. This is particularly relevant in films, when the hero must banish the monster through death or other means. As outlined in her research, Needy waits until Jennifer is in a feeble state. She then kills her with a stab to the heart. Unfortunately for Needy, she is bitten by Jennifer in the resulting struggle, so, in the tradition of vampirism, the film ends with Needy becoming a succubus herself.

As we delved into the research of succubi, we became fascinated by the rarity of female monsters. According to the United States Department of Justice, 89.5% of convicted perpetrators of homicide are male.[2] This led us to dig deeper into female killers, especially those who killed serially. As they constitute such a stark minority, what are the psychological imperatives of female murderers? And are their motives different than their male counterparts? In order to learn more, we consulted academic research on the similarities and contrasts of male and female serial killers. In a 2015 study published in *The Journal of Forensic Psychiatry & Psychology*, a robust group of sixty-four American female serial killers were analyzed to further our scientific understanding of their psychology. An interesting fact noted in the study seems to support this notion of women using their sexual appeal much like a succubus: "Where data were available, two-thirds of these women had average to above-average attractiveness. This may be a quality the female serial killer uses to her advantage."[3]

The motives of the sixty-four female serial killers were grouped into several categories. The highest number of these women were found to be in the "hedonistic" classification. "Consistent with previous research, we found that these women killed for money, power, revenge, and even notoriety and excitement." Money has long been a motive for female serial killers. At the turn of the twentieth century, plain yet charming Belle Gunness led a number of men to their deaths. Much like the succubus myth, Gunness used her femininity to entrap her prey. Through correspondence she convinced many a farmer to move to her farm in La Porte, Indiana. When the men arrived, cash and valuables in hand, Gunness would murder them and bury their corpses in her fertile farmland. While we can never know what sort of thrill Gunness got from killing, we do know that she gained a hearty pile of money from her deeds. "Baby farmers," women who killed unwanted infants, also murdered for financial comfort. Amelia Dyer is one such murderess who is known as one of the most prolific serial killers in history as she murdered between two hundred and four hundred children. Dorothea Puente, who killed the elderly and mentally disabled so she could collect on their social security, is another example of a hedonistic serial killer.

Men, on the other hand, tend to kill serially for more deviant reasons. As outlined in a 1995 study, their motivations are more sexually sadistic: "it appears that a substantial proportion of male serial murderers violate their victims sexually, it is important to examine the role sexual behavior has in the killings."[4] A sexual impetus in female perpetrators of serial murder is almost non-existent. While females tend to use their sexuality like a succubus, leading men like the Pied Piper to their eventual death, there are very few examples of women admitting to sexual thrill as a reason for murder. One example of this deviation from the rule is killer nurse Jane Toppan, who described to police that she got an "erotic charge"[5] from climbing into bed with her dying patients who she had killed with a fatal cocktail of medicine.

The aforementioned 2015 study in *The Journal of Forensic Psychiatry and Psychology* further contends that this difference in female and male serial killers is most probably evolutionary:

The fact that such women primarily kill for resources and such men primarily kill for sex follows evolutionary prediction of sex-specific fitness maximization tactics based on differential reproductive potential. That is, due to differential reproductive potential (i.e., unlimited sperm production vs. very limited ova), in the environment of evolutionary adaptedness, it would have been reproductively beneficial for men to seek multiple sexual opportunities and for women to seek a stable, committed partner with sufficient resources. Evidence suggests that men and women worldwide still seek mates according to this strategy. That fact that male serial killers typically commit their crimes for sex and female serial killers typically commit their crimes for money thus follows evolutionary theory.

It's strange to think that the biological factors differentiating males from females are also what spurn our warped reasons to kill. Female serial murderers are oddly motivated by the same desires as other women, the evolutionary need to procure resources. It seems important to note that nearly all the serial murderers named in this chapter were attempting to function in a misogynistic society on the cusp of women's suffrage. Though their acts are despicable and unforgivable, this could provide further insight into why women have killed.

Biological factors affect people's reasons for committing murder.

Uniquely female, the succubus legend shares aspects with the vampire. Succubi like Jennifer in *Jennifer's Body* employ their charm, just as attractive male vampires in movies like *Interview with the Vampire* (1994) capitalize on their sex appeal. And then, once the prey is properly spellbound, they are devoured for the monster's own gain. It was what real female monsters hoped to gain that led us to find that evolution was a key player in the formation of motive. In proof, we only have to go back to that seemingly pleasant farm in La Porte, Indiana, where the bodies of men languished beneath the soil. During her reign of horror, Belle Gunness, coined "Lady Bluebeard" placed an ad in a local paper: "comely widow who owns a large farm in one of the finest districts in La Porte County, Indiana, desires to make the acquaintance of a gentleman equally well provided, with view of joining fortunes."[6]

SECTION FOUR

REANIMATED CORPSES

CHAPTER TEN

NIGHT OF THE LIVING DEAD

Title: Night of the Living Dead	
Year of Release: 1968	
Director: George A. Romero	
Writer: John Russo, George A. Romero	
Starring: Judith O'Dea, Duane Jones	
Budget: $114,000	
Box Office: $30 million	

Many agree that it was in the fall of 1968 when zombies were born. *Night of the Living Dead* premiered as a Saturday matinee at the Fulton Theatre in Pittsburgh on October 1st. This first showing, mostly attended by teenagers and their tagalong siblings, made cinematic history by ushering in the now ubiquitous depiction of ambling, brain-ravenous zombies. When actor Bill Hinzman playing the part of "Ghoul" stumbled into the graveyard to Tom's (Keith Wayne) famous utterance, "They're coming to get you, Barbara!" Hinzman unknowingly became the first modern-day zombie. His performance, replete with a stiff-legged walk, soulless eyes, and silent, twisted mouth, informed the portrayal of the undead for the succeeding five decades.

Reanimated corpses were not exactly new to film when *Night of the Living Dead* premiered in the late 1960s. *The Plague of Zombies* was released by Hammer Pictures in 1966, and the representation of a somnambulist or sleepwalker first appeared on celluloid in the German picture *The Cabinet of Dr. Caligari* (1920). *Caligari* paralleled Haitian zombie lore with an individual under someone else's control. Although not technically a zombie, they had a lumbering gait and lack of cognitive ability. But it is the visual performances and tropes, like humans being trapped by a horde of zombies, borne from *Night of the Living Dead*, that have spawned everything from the movie *Night of the Comet* (1984) to the television series *The Walking Dead* (2010–present).

While George Romero refined the zombie genre in film, the concept of the undead has existed in numerous cultures for ages. References to zombie-like creatures date back as far as the writings of Gilgamesh in 2100 BCE. *The Epic of Gilgamesh* from ancient Mesopotamia is considered to be the oldest surviving work of literature. It contains the haunting warning "the dead go up to eat the living! And the dead will outnumber the living!" This is strikingly similar to the well-known quote from 1978's *Dawn of the Dead*, "when there's no room in Hell, the dead will walk the Earth!" In China, the undead were known as the *jiang shi*. These creatures killed people in order to steal their life force, or *qi*. This lore dates back to the Qing Dynasty and the scholar Ji Xiaolan. He cited various reasons for bodies coming back to life including possession or a person's soul not leaving their body. There was even the belief that if a pregnant cat leapt across your coffin you would be zombified! In Scandinavia the myth of *draugr* dates back to the eighth century. The *draugr* were believed to rise up from the dead to guard the treasures in their graves. And in twelfth century England, regarded historian William Newburgh wrote of "corpses [that] come out of their graves."

Most modern horror movies trace their version of the undead to folklore revolving around the religion of voodoo in the country of Haiti. Voodoo folklore views zombies as bodies without souls. For one to become a zombie, "zombification" must occur. A sorcerer, or *boko*, performs a spell on a person to kill, enslave, or sicken them. They may perhaps use poisonous powders such as frog or toad venom and tetrodotoxin, a powerful neurotoxin that is secreted by puffer fish. This toxin can trigger paralysis or death-like symptoms and cause others to believe the person is dead. Other steps in the zombification process include keeping the *ti-bon anj*, the manifestation of awareness and memory, in a special bottle. The zombie will remain a slave to the sorcerer until the bottle is broken or the zombie ingests salt or meat.

In 2014, George Romero spoke to NPR's Arun Rath about the inspiration behind his zombie franchise.[1] He explained that he "grew up on classic movie monsters" and never dreamed he would be the father of the modern-day zombie, "I never expected it. I really didn't, all I did was I took them out of 'exotica' and I made them the neighbors ... I thought there's nothing scarier than the neighbors!" Although Romero's film went on to earn $12 million at the box office (two hundred and fifty times its modest budget) years before Kelly and I (Meg) were born, both of us have had our own personal experience with this groundbreaking horror film. Kelly remembers a Christmas morning when she unwrapped a VHS copy of *Night of the Living Dead*, a special gift from her father. It was the first horror movie she ever watched, ultimately the catalyst in sparking her adoration for all things spooky. Meg fell in love with the film's sequel, *Dawn of the Dead*, before she laid eyes on the original. *Dawn of the Dead* holds much of the same charm as its predecessor, yet with a bigger budget, therefore a more ambitious representation of zombie carnage. A decade later, Romero was able to add a greenish pallor to his zombies as well as a neon red blood to the victims. This undoubtedly upped the horror ante.

More than fifty years later, the legacy of *Night of the Living Dead* continues. Zombies have been inculcated into the cultural zeitgeist alongside vampires and slashers. The concept of reanimation, and the ultimate betrayal of a family member or friend coming back from the dead to snack on your brain, still terrifies. Although, now that many of us have become more sophisticated in our viewership, quite a few have come to complain about the inconsistencies of body decomposition in zombie films and television. One target of this social media murmur is the aforementioned AMC hit *The Walking Dead*. Fans on Twitter and beyond began to ask if the show was accurately representing how a body would decompose over time. In 2015, MTV News even interviewed forensic anthropologist, Kimberlee Moran, to explain the body decomposition on the popular TV series. Moran describes that the zombies on the series "would have all kinds of parts of their body dropping off of them all the time, until they become a skeleton."[2] To further our understanding of how real bodies would act under the extreme and thankfully unreal affliction of zombification, we interviewed a medical expert about death and the body decomposition in

George Romero's first two films—someone Meg and I (Kelly) both know (who hides his eyes while watching horror movies and even screams at the frightening scenes!)—Meg's husband, Dr. Luke Hafdahl, an internal medicine physician at the world-renowned Mayo Clinic.

Meg: **"In horror films, including** *Night of the Living Dead,* **we see actors 'die.' Can you tell us how a body reacts to death in its immediate stages?"**

Dr. Hafdahl: "Generally speaking, when someone is dying, there are two roads to death. Most people simply become withdrawn, lethargic, and eventually comatose. They stop speaking. Their breathing becomes altered and, often, you can hear a sound called the 'death rattle,' a rattling sound in the chest that comes from vibrating secretions. People's fingers become blue from lack of blood flow.

The other road to death is less common but much more dramatic, called 'terminal delirium,' in which people become agitated, confused, hallucinate, restless, and begin having involuntary muscle jerks. When people die, the color immediately drains from their face and they develop an ashen appearance. Still, it can be hard to judge if they are alive without checking for a pulse."

Meg: **"Well, I guess that explains that pale zombie pallor!"**

Kelly: **"Are there any diseases or conditions that make someone seem as if they are dead?"**

Dr. Hafdahl: "One of the more terrifying syndromes is called pseudo-coma or locked-in syndrome in which someone has a stroke in their brainstem in such a way that their consciousness is preserved (they are completely aware of everything) but they lose their ability to move their limbs, speak, and swallow. They can communicate only by blinking or moving their eyes up and down (they often cannot move their eyes from side to side)."

Kelly: "That *is* zombie like, not having control over your own body."

Meg: "What about a sort of partial death? Zombies seem both alive and dead."

Dr. Hafdahl: "Certainly, you can see a spectrum of death. People can have parts of the body die before the brain, like people with peripheral vascular disease where their blood vessels are too narrow to deliver blood and nutrients, so their limbs essentially die and need to be amputated (gangrene). Strokes are essentially partial brain death, where a part of the brain dies from a lack of blood flow, so whatever is controlled by that part of the brain dies (i.e., if they have a stroke in the speech center, they have slurred speech, if they have a stroke in the motor cortex, they develop weakness in half of their body, etc.)."

Kelly: "In *Night of the Living Dead* 'fresh' zombies have pale skin, walk stiffly, and sometimes moan. As a man of science, what advice would you have given George Romero about how a zombie should move or act?"

Dr. Hafdahl: "Think of living things like a car on the highway. In addition to the vehicle (the body), it needs two things to work: a driver (the brain) and a fuel source (food). Zombies would be no different! From the brain perspective, I think Romero did a very good job. In order for it to make sense I think the term 'undead' is the most proper term to use as we need to think of zombies partially living. Death is loss of integrated function of various organ systems, so brain death is when we lose that integrated function: the control center (i.e., the brain)

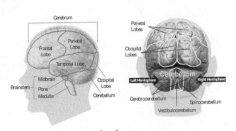

The brain.

can no longer control the body. Thus, if zombies are completely dead, the brain, by definition, could exert no control over the body in any capacity. Zombies would have no volitional movement so they would just lie there, which, of course, would not make for a scary movie. 'Undead' implies that death occurred, but there is some partial return to life, and the brain is able to regain control (at least partially) over the body again. I can imagine that a zombie, which has returned to partial living and has a partially functioning brain, probably has portions of its brain that still work but also parts that are very damaged. In humans, we can see what happens when small portions of the brain die, such as with strokes, multiple sclerosis, and other neurodegenerative diseases. One of the most interesting parts of brain physiology is that everything you do is controlled by a certain part of the brain. If the speech area of the brain dies, people lose the ability to talk and understand language. If the frontal lobe is damaged, people lose the ability to control their impulses, etc. I imagine a zombie has sustained profound damage to much of its brain, which would explain its loss of speech, impulse control, ability to feel pain, balance, etc. which I think Romero has perfectly captured with these stumbling, aphasic, primitive creatures in human bodies. I'm not sure why motor function is so much better preserved than other parts of the brain, but I guess that they have to move to be threatening!"

Meg: **"Do you think that's why Romero chose to kill the zombies with a shot to the head? Because, supposedly, there is one part of the brain still functioning?"**

Dr. Hafdahl: "I would suspect so! It seems to make sense that you kill the undead by severing the 'control' to make them dead again."

Meg: **"Or deader! Is that a word? More dead?"**

Dr. Hafdahl: "Ha, yeah, exactly! So, going back to the car analogy, the other thing that zombies would need to 'live' would be a fuel source (and the ability to use that fuel source). Zombies are no different than any other animal. We need to consume food for energy in order to run the engine, and we all have a tremendous, overpowering primal urge to obtain energy. It makes sense that living humans would be a ripe source for energy! However, I would suspect that zombies would be omnivorous like us (i.e., can eat both animals and plants for food), so I suspect they are probably snapping apples off of trees when there are no humans around."

Kelly: **"That's really cool! I wondered if there was any biological reason zombies would want to eat humans. And it's funny to think of them eating other foods, too. Supposing there was a parasite or disease that brought upon zombies, how long would we have until the zombies crumbled? Could weeks or months go by like in the case of *Dawn of the Dead*?"**

Dr. Hafdahl: "Great question. It would depend on answering this: could zombies actually digest humans for fuel? A human body can go about three weeks without food, so you would assume the same is true for zombies if they are following our rules. In order to stay alive, we need to turn the food we eat into a form of fuel we can use (glucose) through a process called digestion. This is a very complex process that involves a lot of things to work, including a stomach and at least the first part of the small intestine. Decapitated zombies would definitely be in trouble since they would need a mouth, but I think most zombies could still do this. Zombies would then need to be able to burn the fuel for energy. Our bodies do this through aerobic respiration, in which glucose and oxygen are combined to release energy. That energy is used to make muscles contract, keep skin intact, make the heart pump, etc. However, one glance at a zombie tells us that this would not happen. Despite their decomposing skin and no need for working hearts or lungs, most zombies are clearly 'living' longer than three weeks (though, if I recall, the film *28 Days Later* (2003) adheres to this rule). Thus, zombies clearly must have some other process that lets them harness energy, or at least slows down the decaying process. Perhaps part of the science of the zombie virus is that it allows them to make energy without oxygen?"

Meg: **"Hmm, so it sounds like zombies might be starving, giving us maybe a few weeks until they die."**

Kelly: **"Again!"**

A natural and ultimately universal process, death is our greatest mystery. The zombie apocalypse may not be coming anytime soon, but it has been fascinating learning about the biology of death. Science can aid our understanding of the concept of zombies, but more importantly teach us about our own biological demise. Thanks to Meg's husband for chatting with us. He may be too afraid to watch horror movies, but he sure does know a lot about death!

It might be terrifying to think of our loved ones biting into our flesh, although unlike their monster counterparts, zombies are not malicious. George Romero describes this fundamental difference, "you can't really get angry at them, they have no hidden agendas, they are what they are."[3] Perhaps there is something comforting about the prospect of coming back from the dead? If only to stumble mindlessly in search of a delicious, screaming meal.

CHAPTER ELEVEN

FRANKENSTEIN

Year of Release: 1931
Director: James Whale
Writer: Francis Edward Faragoh, Garrett Fort
Starring: Boris Karloff, Colin Clive
Budget: $262,000
Box Office: $12 million

"It was the secrets of heaven and earth that I desired to learn; and whether it was the outward substance of things or the inner spirit of nature and the mysterious soul of man that occupied me, still my inquiries were directed to the metaphysical, or in its highest sense, the physical secrets of the world."[1] This quote from Mary Shelley's novel *Frankenstein* (1818) encapsulates our shared pursuit of knowledge. The creature, not unlike his creator, desires to understand this complicated and mysterious planet we all live on. *Frankenstein* is a novel about science, about its magnificence, but more, about its danger to our society and to our souls. If Mary Shelley lived today, we would have to assume *Frankenstein* would be a treatise on technology. The creature, borne of our collective pride and stubbornness, would perhaps come in the form of an iPhone.

But in 1818, the nineteen-year-old Shelley, often credited as one of the first horror storytellers, focused instead on the tenuous science of medicine. Her Dr. Frankenstein actively rejects domesticity and convention in order to pursue scientific glory. And, once he achieves his goal, he rejects the creature he has created. *Frankenstein* is a cautionary tale. It is horror at its finest, exposing the ethical dilemma memorably summed up by Dr.

Ian Malcolm (Jeff Goldblum) in the sci-fi horror film *Jurassic Park* (1993) when faced with scientifically engineered dinosaurs: "your scientists were so preoccupied with whether or not they could that they didn't stop to think if they should." In the Industrial Age, when resurrectionists were digging up the recently deceased for medical dissection, Mary Shelley was the first to ask in literature if ethics and science could coexist. Her novel, rife with meaning, is also terrifying in its vivid accuracy of the runaway train that is modern science and technology.

Popular in its time, *Frankenstein* was adapted for the stage by Richard Brinsley Peake. *Presumption; or, the Fate of Frankenstein* (1823) was attended by Shelley herself at the English Opera House (now known as the Lyceum Theatre) on the West Side of London. Yet another play, *The Man and the Monster!; Or, the Fate of Frankenstein* played at the Royal Coburg Theatre in London in 1826, and sixty years later the creature was revived for a musical burlesque entitled *Frankenstein, or the Vampire's Victim* (1887). In 1910, the legend of Dr. Frankenstein and his ill-begotten creation came to the screen for the first time. The twelve-minute movie, produced by Thomas Edison's Edison Studios, was shot in a few days by prolific director J. Searle Dawley. The film downplayed many of the macabre elements of Shelley's masterpiece:

> In making the film the Edison Co. has carefully tried to eliminate all actual repulsive situations and to concentrate its endeavors upon the mystic and psychological problems that are to be found in this weird tale. Whenever, therefore, the film differs from the original story it is purely with the idea of elimination of what would be repulsive to a moving picture audience.[2]

Imagine if these constraints were still put on horror films! After the Italian silent adaptation, *The Monster of Frankenstein* (1920), Americans at Universal Studios once again brought the creature back from the dead for the most iconic depiction to date. *Frankenstein* (1931) is a classic horror picture, one so certain of its scares, it included a warning to audiences that "it will thrill you. It may shock you. It might even *horrify* you." Bela Lugosi, who had recently donned Dracula's cloak to much avail, famously quit *Frankenstein* when he was underwhelmed by the test makeup. This choice to leave, and to allow Boris Karloff to take his place as the creature, has long been considered Lugosi's biggest career downfall. The fallout of this lofty mistake is portrayed in the Tim Burton film *Ed Wood* (1994).

The 1931 *Frankenstein* has been the touchstone for later adaptations of how the creature looks, moves, and acts. Unlike the thoughtful, empathetic creature of Shelley's creation, the film's monster appears as slow-witted, and as more the villain than Dr. Frankenstein. Universal Studios also brought back Igor, the doctor's assistant, who had not been in the novel, but had first been portrayed in Peake's 1823 play adaptation. While Universal's version of *Frankenstein* may seem tame by our modern standards, there were controversial scenes that unnerved many. The most contentious is when the creature throws little Maria (Marilyn Harris) into the water, accidentally drowning her. State censorship boards in Massachusetts, Pennsylvania, and New York all cut this scene before allowing moviegoers to experience *Frankenstein*. Despite some grumblings about decency, *Frankenstein* went on to be both a financial and critical success. In 1991 it was added to the National Film Registry, and was also eighty-seventh on AFI's 100 years...100 movies. *Frankenstein* endures. It is, perhaps, the best example of when horror and science intersect.

In *Frankenstein*, Dr. Frankenstein is able to piece together a body made of several different bodies' parts. Is this scientifically possible? Although reanimating a pieced-together corpse may not be realistic, limb transplants are no longer science fiction. Hands and arms have successfully been transplanted and the first leg transplantation may not be far behind. How does the process work? Organ donation is only possible if the organ in question has blood and oxygen flowing through it until the time of harvesting. A living donor can give a whole kidney, a portion of their liver, lung, intestine, or pancreas.[3] For a hand transplant the donated hand usually comes from a brain-dead donor. The procedure to connect the new hand can take anywhere from eight to twelve hours. The recipient then needs to take immunosuppressive drugs in order to prevent rejection of the hand and take part in physical therapy to gain function and mobility. The first hand transplant took place in 1964, and in 2016 the first double arm transplant took place. The recipient, a veteran who was a quadruple amputee, has been able to gain the use of both arms and even threw out the first pitch at a baseball game in 2018.

Every day thousands of people are the recipients of donated organs. In the United States in 2017 more than 34,000 people received organs such as corneas, kidneys, and hearts. Although that number is impressive, there are more than 114,000 people on waiting lists for organs, and twenty or more die each day by not receiving one in time. This need has led to an increase in the black market for organs. Illegally harvesting organs is a dark reality that several horror movies have focused on, including *Turistas* (2006) in which a group of friends on vacation in Brazil get caught in an organ harvesting trap. How can organs possibly be harvested? Some "organ brokers" may have connections with funeral homes in order to get organs before a body is embalmed or cremated. Some people are willing to part with their own organs for financial

Organ transplants occur every day.

compensation while others may have organs taken from them against their will. According to the World Health Organization, approximately seven thousand kidneys are illegally harvested annually by traffickers worldwide with the average buyer paying $150,000. Several cases over the past decade prove how the black market for organs is alive and well. In China, a missing six-year-old boy was found alone in a field. Both of his eyes had been removed, presumably for the corneas. In 2012, a young African girl was kidnapped and brought to the UK for the sole purpose of harvesting her organs. She was rescued before any procedure was performed. Kendrick Johnson, a Georgia teen, died at school in January of 2013 under mysterious circumstances. After his parents obtained a court order to have the body exhumed for an independent autopsy, the pathologist found the corpse stuffed with newspaper and the brain, heart, lungs, and liver missing, leading some to believe he was murdered and his organs sold on the black market.

There are plenty of horror movies that portray organ transplants gone wrong. *The Eye* (2002) and *Body Parts* (1991) each explore what happens when a person receives more than just the organ after surgery. In *The Eye*, a blind woman receives a cornea transplant and regains her sense of sight. She also gains the ability to see the dead and deaths foretold around her. It makes for an unsettling and creepy plot. The writers said they were inspired by a report they had seen in a Hong Kong newspaper about a sixteen-year-old girl who had received a corneal transplant and committed suicide soon after. "We'd always wondered what the girl saw when she regained her eyesight finally and what actually made her want to end her life."[4] In Wes Craven's movie, *Body Parts*, a detective is given the arm of a convicted serial killer who was sentenced to death. He begins to envision the murders the other man committed and begins to act violently himself. Art hit a little too close to home at the time of the film's release in Milwaukee, Wisconsin. Paramount Pictures pulled ads for *Body Parts* after police found dismembered bodies in Jeffrey Dahmer's apartment.

Recipients of limbs may experience phantom limb pain after losing their own body parts. How is this possible? Doctors still have no clear consensus as to its cause, but many think it results from changes in the peripheral nervous system. There are different sensations that amputees may feel after their amputation. A person experiencing "telescoping" has the feeling their missing limb is still there, but that it has shrunk to a very small size, similar to a collapsed telescope. It may not be painful but it is unnerving. Phantom pain has amputees reporting a physical sensation of pain in their missing limb. Even though amputations have occurred throughout history, phantom pain became more prevalent, or documented, during the Civil War. A physician during that time noted that 90 percent of amputees reported phantom limb pain. Can anything be done for this condition? One of the most effective therapies is mirror box therapy. The patient watches in a mirror while receiving physical therapy in order to remap the brain's neural pathways to register that the limb is no longer there. Other treatments include medication or injections to help alleviate the pain.

Although modern day scientists may not be putting together Frankenstein-like creatures in their labs, we are closer than ever to seeing the science fiction aspects from Mary Shelley's novel become a scientific reality.

CHAPTER TWELVE

THE MUMMY

Year of Release: 1932	
Director: Karl Freund	
Writer: John L. Balderston	
Starring: Boris Karloff, Zita Johann	
Budget: $196,000	
Box Office: $3 million	

We all have the image of a bandage-wrapped mummy lumbering about after emerging from its sarcophagus. But when did mummies, in the sense that we know them, become prevalent in pop culture and media? One of the earliest examples of undead mummies is *The Mummy!: Or a Tale of the Twenty-Second Century*, a novel written by Jane C. Loudon in 1827. This early science-fiction work is about an Egyptian mummy named Cheops, who is brought back to life in the twenty-second century. Another woman, Louisa May Alcott, wrote *Lost in a Pyramid; or, the Mummy's Curse* in 1869 in which some seeds found in a tomb bring back the curse of sickness and early death.

What was the inspiration for the Universal Studios monster, *The Mummy*? The 1922 discovery of Tutankhamun's tomb received worldwide press coverage. The tomb was nearly intact and allowed the world to see what a sarcophagus looked like. It sparked a renewed public interest in ancient Egypt and inspired film producer Carl Laemmle Jr. to find a story similar to *Dracula* or *Frankenstein* but based on Egyptian horror. The story behind *The Mummy* resembles Sir Arthur Conan Doyle's *The Ring of Thoth* (1890) in which a mummy is resurrected in a museum. Jack Pierce, a

Hollywood makeup artist who also created the iconic look for *Frankenstein*, was brought on board to transform actor Boris Karloff at 11 a.m. on the day they filmed *The Mummy's* opening sequence. Karloff's makeup consisted of cotton, collodion, and spirit gum. He was then wrapped in bandages treated with acid and burnt in an oven. Filming began at 7 p.m. and ended at 2 a.m., followed by nearly two hours to remove all of the makeup. Karloff found the removal of gum from his face painful, and overall found the day "the most trying ordeal I [had] ever endured."[1]

Boris Karloff's character in *The Mummy*, Imhotep, is discovered wrapped like a mummy but having been buried alive. He is brought back to life by the reading of a scroll and sets out to find his lost love. Are there cases of people actually being buried alive? Several Edgar Allan Poe stories explore this fear including *The Premature Burial* (1844), *The Fall of the House of Usher* (1839), and *The Cask of Amontillado* (1846). The chance of being buried alive was a real threat prior to modern medicine. With the absence of scientific certainty, the public, and even medical professionals, were less able to determine if a person was actually dead. Some tests were done on the bodies to see if the person would wake up. These included pinching the person's nipples and even inserting a hot poker into the rectum. No, thank you! A famous case of premature burial took place in 1891 when a girl by the name of Octavia Smith was presumed dead and buried after suffering from a mysterious illness. A few days later several others in the community came down with a similar illness. It was discovered that their shallow breathing was caused by the tsetse fly, which causes African Sleeping Sickness, characterized by symptoms of extreme lethargy. Octavia's grave was dug up to reveal a terrifying sight: scratch marks lining the coffin, her hands bloody, and a look of terror on her dead face.

Taphephobia, the fear of being buried alive, led to some interesting inventions. Safety coffins had several features to put people's minds at ease: glass tops for observation, ropes connected to bells in the graveyard to signal passersby, and even breathing tubes to allow for plenty of air while the person buried alive waited to be rescued. Even though these features were used for decades, there are no known people who were saved by these inventions from being buried alive.[2] Boris Karloff's character in *The Mummy* could have used an alert system to let others know he was wrapped and buried alive.

Ancient Egyptians believed in an immortal soul and thought that embalming, the act of temporarily preserving a body for funeral or transport, preserved the body so that the soul could someday return. The science of embalming began over three thousand years ago in Egypt. Bodies in Egypt decomposed quickly due to heat and bacteria. The mummification process allowed for bodies to be preserved and not disintegrate. There are five steps in the process:[3]

1. **Remove the brain:** this was done through the nostrils with special hooks.
2. **Remove the internal organs:** these were removed through the chest and then placed into special jars to be buried with the mummy.
3. **Immerse the body in salt:** this step thoroughly dried the body.
4. **Dehydrate the body in the sun:** this step got rid of any moisture left.
5. **Wrap the body in bandages:** hundreds of yards of bandages were needed to wrap a single body.

The mummification process would take up to seventy days to complete. Because of the elaborate process, mummification was expensive. Although some common people were buried like mummies, it was mostly prevalent for pharaohs and members of nobility. That being said, it's interesting to note that some animals were also mummified, especially if they were considered to have religious significance. There are even cases of "self-mummifying" found in Buddhist culture. *Sokushinbutsu* is the practice of reducing food and water intake to eventually die and become a mummy.

A depiction of the mummification process in ancient Egypt.

How were bodies treated for burial in other cultures? In Persia, bodies were placed on a high rock and left for birds and dogs to devour. After the bones had been stripped, they were buried in a pit. Ancient Mayans buried their dead with corn in their mouths as a symbol of rebirth. Romans believed in keeping the living and the dead separate and only buried bodies outside of the city walls while ancient Greeks believed the afterlife existed underground.[4]

When did embalming become common practice? The anatomy acts of 1832 and 1883 allowed medical professionals to make many medical and scientific discoveries. Previous to this law, only the corpses of executed murderers could be used for the study of anatomy. Some medical schools were resorting to grave robbery to learn about and practice dissection. By allowing the dead to be used more readily, advancements were made in how to embalm bodies. Embalming bodies, as we do today, came into common practice during the Civil War.[5]

Many horror movies have scenes that take place in funeral homes or mortuaries. *I'm Not a Serial Killer* (2016) features several scenes of embalming and funeral home work. The Netflix series *The Haunting of Hill House* (2018) explores death and how the viewing of a deceased loved one could be cathartic. What are the current scientific practices in embalming bodies or preparing them for burial? The first step is to wash the body in a disinfectant solution. Limbs are massaged to relieve the stiffening of the joints and muscles and the body's eyes are closed using glue or plastic eye caps. The lower jaw is secured by wires or sewing and can be manipulated into a desired position. During the surgical portion of the embalming process, blood is removed from the body through the veins and replaced with formaldehyde-based chemicals. Next, the body's cavities are embalmed. A small incision is made in the lower part of the abdomen and the organs in the chest cavity and abdomen are punctured and drained of gas and fluid contents. Once the incision is closed, the process is complete. For those choosing an open casket for their funeral, hair is washed and styled and makeup is applied. How did or do people around the world hold funerals for the dead? In some cultures, it was thought that the best way to honor the dead was to eat them. Endocannibalism was practiced in parts of Papua New Guinea and Brazil and thought to create a permanent connection between the living and the dead. In funeral practices of the Marinoa people of Australia, corpses were left to decompose and the liquid from the body was collected. The thought was that the good qualities of the person who died would be passed on through this liquid so it was rubbed on others.[6]

More modern funeral and burial practices include cremation, turning the deceased's body into beads, and other environmentally friendly techniques. Some people are choosing to skip embalming and instead be buried in biodegradable coffins. Whatever you choose to do with your remains, hopefully you can rest peacefully knowing that scientific advances have only helped our chances of avoiding problems that *The Mummy* had.

SECTION FIVE
THE POSSESSED

CHAPTER THIRTEEN

THE EXORCIST

Year of Release: 1973	
Director: William Friedkin	
Writer: William Peter Blatty	
Starring: Ellen Burstyn, Linda Blair	
Budget: $12 million	
Box Office: $441.3 million	

In 1973, audiences lined up in droves to see *The Exorcist*. Footage from that era shows people excited to enter the movie theater, akin to Black Friday shoppers, with smiles on their faces. The footage of audiences exiting tells a different story; people look noticeably freaked out. Some even reportedly fainted or vomited during the movie. Whether you saw the iconic movie in the theater when it first came out or years later, it certainly leaves a lasting impression. *The Exorcist* was based on a novel of the same name by William Peter Blatty, which came to life after Blatty read an article about a fourteen-year-old boy who was possessed by demons. The article stated that a priest performed an exorcism and the boy was able to go on with his life. The idea stuck with Blatty and the novel (1971) was written after years of research. He was able to talk to the priest who performed the exorcism and changed some details in order to protect the boy's anonymity.

The exorcism of Robbie Doe, or Roland, has been investigated and refuted by some but the facts remain. In 1949, a boy was hospitalized and more than one priest performed the exorcism ritual on him. The story begins with the family of the boy reporting strange happenings in their home: objects flying around the boy, his bed shaking, and scratching noises in the walls. Their local pastor put them in touch with a priest who claimed to have witnessed the same phenomenon when around the boy. On August 19th, 1949, *The Evening Star* in Washington, D.C., featured the article "Priest Freed Boy of Possession by Devil, Church Sources Say."[1] The article opens by saying, "A Catholic priest has successfully freed a fourteen-year-old Mount Rainier, Maryland, boy of reported possession by the devil here early this year, it was disclosed today." The article also states that the boy was studied at both Georgetown University Hospital and St. Louis University.

What really happened to Robbie Doe? According to the book *Possessed: The True Story of an Exorcism*,[2] author Thomas B. Allen says "the consensus of today's experts" is that "Robbie was just a deeply disturbed boy, nothing supernatural about him." Another writer, Mark Opsasnick, states "Roland Doe was simply a spoiled, disturbed bully who threw deliberate tantrums to get attention or to get out of school."[3] Religious experts insist that true demonic possession cannot be explained by psychiatry. Whether this case was true or not, it certainly inspired *The Exorcist* novel and film, which sparked a large public interest in exorcisms and Catholicism.

Are there other documented examples of exorcisms? The exorcism of Anneliese Michel is another well-known case which served as the basis for the film *The Exorcism of Emily Rose* (2005). When Anneliese was sixteen years old she suffered a seizure and was later diagnosed with temporal lobe epilepsy, a chronic condition. The seizures associated with this disorder can cause hallucinations, amnesia, and unprovoked fear or anxiety. She was later diagnosed with depression and spent years being treated with medication. After several years, Anneliese developed an aversion to religious objects and claimed to hear voices. She showed aggressive traits, harmed herself, and ate insects while hospitalized. She and her family became convinced that she was possessed by a demon and appealed to the Catholic Church to have an exorcism performed on her. In 1975, permission was granted and exorcism sessions were conducted by two priests over the course of ten months. In 1976, Anneliese died of malnutrition and dehydration. Her autopsy found that she weighed only sixty-eight pounds and had pneumonia at the time of her death.

The priests and her parents were all found guilty of negligent homicide, a charge brought against those who knowingly allow someone else to die when it could have been prevented, and were sentenced to six months in jail.

There are other, more recent examples of exorcisms gone wrong. In 2003, an autistic eight-year-old boy in Milwaukee, Wisconsin, was killed during an exorcism by church members who blamed a demonic possession for his disability. In 2005, a young nun in Romania died at the hands of a priest during an exorcism after being bound to a cross, gagged, and left for days without food or water in an effort to expel demons. In 2010, a fourteen-year-old boy in London, England, was beaten and drowned to death by relatives trying to exorcise an evil spirit from him.

Can the symptoms of possession be medically explained? Several disorders have been associated with possession including Tourette's syndrome and schizophrenia because of symptomatic erratic and psychotic behavior. Possession can also be linked to dissociative identity disorder with about 29 percent of those with the disorder identifying themselves as demons. Many mental and mood disorders could cause the psychological symptoms of possession.

Besides the mental health aspect, there are some physical presentations of being possessed. Some victims of possession report words appearing on their skin. Doctors believe this can be explained by dermatographic urticaria, a disorder that translates to "writing on the skin." Those with this disease can create welt-like lines on their skin just by applying some pressure. This could appear quite worrying to anyone with no knowledge of the condition. Another physical symptom in possession cases is the vomiting of objects. This could be explained by an eating disorder known as pica. People who suffer from this disorder are known to eat non-nutritive things such as dirt, glass, and other items.

What does an exorcism entail? According to the Catholic Church:

The priest delegated by the Ordinary to perform this office should first go to confession or at least elicit an act of contrition, and, if convenient, offer the holy Sacrifice of the Mass, and implore God's help in other fervent prayers. He vests in surplice and purple stole.

Having before him the person possessed (who should be bound if there is any danger), he traces the sign of the cross over him, over himself, and the bystanders, and then sprinkles all of them with holy water. After this he kneels and says the Litany of the Saints, exclusive of the prayers which follow it. All present are to make the responses.[4]

Malachi Martin, a former Jesuit priest and self-proclaimed exorcist, offers additional information in the book *Hostage to the Devil*.[5] He considers there to be four stages of an exorcism:

1. **Pretense:** The demon is hiding its true identity.
2. **Breakpoint:** The demon reveals itself.
3. **Clash:** The exorcist and the demon fight for the soul of the possessed.
4. **Expulsion:** If the exorcist wins the battle, the demon leaves the body of the possessed.

Does the Catholic Church endorse or train exorcists? The Vatican first issued official guidelines on exorcism in 1614 and revised them in 1999. According to the U.S. Conference of Catholic Bishops, signs of demonic possession include superhuman strength, aversion to holy water, and the ability to speak in unknown languages. Other potential signs of demonic possession include spitting, cursing, and "excessive masturbation." There is an official International Association of Exorcists, which represents more than two hundred Catholic, Anglican, and Orthodox priests. The Vatican itself held an exorcist training workshop in 2018 after the increase of reported demonic possessions. The Church of England released guidelines that say "doctors, psychologists, and psychiatrists should be consulted where appropriate,

The crucifix is a strong religious symbol in the Catholic faith.

and that deliverance should be followed up with continuing pastoral care and should be done with a minimum of publicity."[6]

Are there stories of demonic possession in other cultures or religions? Anthropologists have concluded that in some cultures, those with little or no social influence can vent their true feelings toward the more powerful members of their society while "possessed" without having to face any repercussions. They are not held responsible for their actions; the possessing spirit is. Also, historically in Europe, it was women who were much more likely to be "possessed" than men. Exorcism rituals are found in many religions around the world, including Hinduism, Buddhism, Taoism, Shintoism, Judaism, and Islam. There are several references to possession by evil spirits and exorcism in the Qur'an and the Bible. Experts agree that religious beliefs can play a role in the diagnosis and treatment of "possessed" people.

One famous scene in *The Exorcist* shows Linda Blair's character Regan projectile vomiting onto the priest. For filming, real pea soup was used and mixed with oatmeal to look more authentic. (Gross!) What is the science behind projectile vomiting? The forced ejection of stomach contents can by caused by a variety of things. In infants, it's most often due to a condition called pyloric stenosis. This condition affects a tube in the child's body that connects the small bowel and the stomach. This condition can be repaired through surgery. In adults, projectile vomiting is quite rare. When it happens though its common causes are food poisoning, toxins such as alcohol, or illness. Although this type of vomiting causes no long-term effects, any type of throwing up can be traumatic.

In some scenes of *The Exorcist* we see the actors' breath when in Regan's bedroom. To get the effect, four air conditioning units were brought in to cool down the space. How cold does it have to be to see your breath? Here in Minnesota it's a daily occurrence during the winter months. Temperatures need to be below forty-five degrees for the moisture in your breath to become visible. On the set of *The Exorcist* the room was reportedly kept below zero. Linda Blair has stated that she dislikes being cold to this day because of that experience.

The death of Father Karras in the movie entails him falling down some stairs. An actual staircase located in Georgetown was used for filming.

Although they were made of concrete, the stairs were padded with half-inch-thick rubber to film the stuntman. Could a fall down a flight of stairs kill a person? Absolutely. According to the National Safety Council[7] over one thousand people die per year by falling down stairs. In fact, stairs are considered the most dangerous part of our homes, particularly for elderly people. To be safe, it's recommended to have your stairs well lit, not carry bulky objects on difficult stairs, and to make sure staircases are free from items that could potentially trip you. It also helps to not be pushed down a flight of stairs by a demon! There is no data on that science.

What would it take for a medical doctor to call in an exorcist? According to Dr. Luke Hafdahl, he would call a priest if the patient presented with that iconic revolving head. "That's when I'd know it's time to try something outside medical science!" Whether you believe in possession or not, the battle between Heaven and Hell rages on in horror cinema. *The Exorcist* was not the first movie to depict body possession by the devil, and there have been dozens since, including; *The Exorcism of Emily Rose* (2005), *Constantine* (2005), *The Possession* (2012), and *The Possession of Hannah Grace* (2018). Yet, nearly fifty years later, Regan MacNeil's violent and shocking body usurpation remains the gold standard of devil versus priest.

CHAPTER FOURTEEN

THE TINGLER

Year of Release: 1959	
Director: William Castle	
Writer: Robb White	
Starring: Vincent Price, Judith Evelyn	
Budget: $250,000	
Box Office: $2.9 million	

Imagine sitting in a movie theater watching a horror film. You're involved in the plot, feeling a little scared, but then you start to feel something else . . . a tingle. It seems that your whole seat is vibrating. Suddenly a skeleton emerges from below the screen of the theater! The plot of the movie seems to have come to life right there with you. This may sound far-fetched, but it's exactly the kind of tactics that William Castle used to promote his films.

The 1959 film *The Tingler,* starring horror icon Vincent Price, focuses on a parasite that attaches itself to humans and feeds on their fear. The writer of the script, Robb White, was inspired by an encounter with a centipede while living in the British Virgin Islands. He likely saw an Amazonian giant centipede which can grow up to twelve inches in length. These animals are carnivores and feed on anything they can overpower and kill. At least one human death can be attributed to these giant centipedes.[1] A nineteenth-century Tibetan poet warned his fellow Buddhists, "if you enjoy frightening others, you will be reborn as a centipede."[2] The plot of *The Tingler* follows that thinking. The movie posits that these centipede-like creatures are part of our bodies and that they feed on our fear. The more fear we have, the bigger the parasite grows. It produces a tingling sensation, eventually crushing its host's spine. The only way to defeat this dreaded beast is to scream.

Are there examples of creatures that become part of other beings in nature? One such creature is the *Cymothoa exigua*, or tongue-eating louse. It is exactly what it sounds like in that it eats the tongue of its host and replaces it. This is the only known parasite that completely replaces a part of the host animal. More than forty species of fish have been known to be infected with this creature.[3] Cookiecutter sharks were originally known as demon whale-biters and are considered ectoparasites because they

The Cymothoa exigua *becomes the tongue of its host.*

attach themselves to the outside of other animals to feed, leaving neat, round bite marks. Because of their preferred environment these parasites rarely come into contact with humans but have been implicated in several attacks. One noted case involved the victim recovering for nine months from the bite after having to get a skin graft procedure.[4]

Some animals are able to manipulate their hosts from within, causing them to act in self-destructive ways that ultimately benefit the parasite. The females of a Costa Rican wasp lay their eggs on the abdomens of orb spiders. After living off its host for weeks, the wasp larva injects a chemical into the spider that makes it build a strange, new kind of web. The web is meant to support the wasp cocoon. After the wasp hatches it will kill and eat the spider.[5]

There are parasites that specifically target humans including tapeworms, scabies, and certain amoeba. Around 70 percent of parasites are not visible to the human eye. Others, like the malarial parasite, can grow up to thirty meters long. In 2017, a six-foot-long tapeworm was removed from a patient through his mouth. The man had been experiencing stomach pains for over two months. A colonoscopy revealed the tapeworm and an endoscopy confirmed it was curled up in his small intestine. After sedation, the tapeworm was pulled from his mouth intact.[6] The horror movie writes itself!

Could a creature attach itself to a human and kill them like the *The Tingler*? The tingler may not exist, but there have been living things found inside of people. One such case was Sanju Bhagat. He always had a large stomach, but in 1999 it became so swollen that he appeared to be in his third trimester of pregnancy. He was having trouble breathing and was rushed to the hospital. Emergency surgery was performed to remove what they thought would be a tumor. After entering the abdomen, though, it became apparent that Bhagat suffered from fetus in fetu, an extremely rare condition in which one twin can be absorbed by the other while in the womb. The fetus can survive as a parasite past birth by forming an umbilical cord-like structure that takes its twin's blood supply. Often times the unknown twin will grow so large that it starts to harm the host, at which point doctors usually intervene. When surgery was performed on thirty-six-year-old Bhagat he had no idea what was inside of him. Doctors described seeing one limb, then another come out. Developed hands with long fingernails, feet, hair, teeth, and bones emerged and were disconnected from the host. Bhagat went on to make a full recovery.

There have been at least two reported cases of people's mouths becoming "impregnated" by squid sperm. After biting into partially cooked squid, squid sperm implanted itself in the mucous membranes of the victims' mouths. Doctors later removed the living creatures from their tongues, cheeks, and gums. In 2013, a four-year-old boy was found to have a live snail living in his knee. Several weeks earlier he had tripped and fallen at the beach, getting a scrape on his knee. The wound didn't heal and was thought to be infected until, upon closer inspection and a squeeze, a sea snail popped out. The boy ended up keeping the snail as a pet and named him Turbo. Now that's a happy ending!

Spiders have been found living inside of multiple people. One woman reported hearing a scratching noise in her left ear and finally went in to see the doctor when her pain became unbearable. A spider was discovered building a home in her ear canal. After several attempts it was finally removed by being sedated. In 2014, a spider entered a man's stomach through an appendectomy incision. It traveled all the way up to his chest, inside of him, before it was removed by doctors. Another woman reported feeling a crawling sensation inside of her head. Doctors discovered a live cockroach in her skull which had entered through her nose. It was removed but doctors said it "didn't want to come out."

These examples are true horror, but what about perceived danger? William Castle, the director of *The Tingler*, used a gimmick called "Percepto," a vibrating device in some theater chairs which activated during the climax of the film. Before the movie began, he appeared with a message:

> I am William Castle, the director of the motion picture you are about to see. I feel obligated to warn you that some of the sensations—some of the physical reactions which the actors on the screen will feel—will also be experienced, for the first time in motion picture history, by certain members of this audience. I say "certain members" because some people are more sensitive to these mysterious electronic impulses than others. These unfortunate, sensitive people will at times feel a strange, tingling sensation; other people will feel it less strongly. But don't be alarmed—you can protect yourself. At any time, you are conscious of a tingling sensation, you may obtain immediate relief by screaming. Don't be embarrassed about opening your mouth and letting rip with all you've got, because the person in the seat right next to you will probably be screaming too. And remember—a scream at the right time may save your life.

As the tingler was tormenting actors on the screen in the movie theater an announcement would play saying that the tingler was "loose in this theatre!"[7] Chairs would vibrate and hired screamers and fainters, who were planted in certain audiences, would be carried out on stretchers.

Are there more recent gimmicks that horror movie producers have used to scare patrons? There may not be instances of tactics as elaborate as those employed by William Castle, but several companies have used unique ways to market their horror movies. When the remake for the classic horror film *Carrie* came out in 2013, the marketing department focused on the psychic abilities of the title character. Using actors and a special effects team, they created a fake telekinetic event in a New York City coffee shop. Patrons saw a stunt man "levitate" and witnessed "spontaneous" movement of tables, chairs, pictures, and books. To promote *Devil's Due* in 2014, an animatronic "devil baby" hit the streets of New York City to get reactions. The video of people's reactions received more than fifty-four million views on YouTube. Regardless of how horror movies are promoted, it's the movies themselves that leave a lasting impression on us. The science of fear may not be revealed to be a centipede-like tingler on our backs, but the feeling is real. The next time you feel that terror watching a horror movie, don't be afraid to scream!

CHAPTER FIFTEEN

GET OUT

Year of Release: 2017	
Director: Jordan Peele	
Writer: Jordan Peele	
Starring: Daniel Kaluuya, Allison Williams	
Budget: $4.5 million	
Box Office: $255.4 million	

Every so often a horror movie comes along that immediately leaves a profound and changing impact on Hollywood. In February 2017, *Get Out* was released to both critical praise and box-office success. Written and directed by Jordan Peele, who was best known as a comedian, *Get Out* snagged four Academy Award nominations. This is a lofty feat for a horror film. Peele went on to be the first African American in history to win Best Original Screenplay. The film's cultural impact looms large, as it is an exploration of race and class tied up in a scary, and often humorous, bow. Peter Debruge of *Variety* described *Get Out*'s complicated appeal:

> The film's subversive POV challenges the place of white privilege from which most pop culture is conceived. By revealing how the ruling majority gives freedoms, but they can also take them away, Peele seizes upon more than just a terrifying horror-movie premise; he exposes a reality in which African-Americans can never breathe easy.[1]

The plot centers on Chris Washington (Daniel Kalyuua) an African American photographer who is visiting his Caucasian girlfriend's Rose

Armitage (Allison Williams) parents for the first time. Right away, the subtext of a mixed-race relationship is discussed. Peele showcases the inherent prejudices Chris routinely deals with, and how this makes for a more stressful meeting. Rose's parents (Bradley Whitford and Catherine Keener) assure Chris that they are of the modern era, they have no issue with their daughter dating a man of color. While their words are reassuring, there is an obvious schism between what they say and how they act. These subtleties, a familiar obstacle for minorities, is depicted through the lens of a horror movie in order to highlight these realities for those, as Peter Debruge described, as having "white privilege." Chris comes to learn that the suburban neighborhood he has found himself in, seemingly pleasant and perfect, is a dangerous place. The older, rich denizens are stealing the bodies of African Americans to implant their minds and consciousnesses within, a sort of body transfer so that they can live longer. Why? Because missing people of color aren't searched for with the same vigor as white people, Peele argues.

While the theme of race is at the center of *Get Out*, there are exceptionally disturbing horror sequences that, even without subtext, are inarguably memorable. For instance, when the truth of Rose is revealed, that she is a cold-blooded sociopath who collects unsuspecting men with her beauty, and when Chris wakes to the video, fully explaining the medical procedure that is going to send him to the "sunken place." This sunken place is where Chris will live out his life, a sort of diminished part of him who must watch as the person who has taken over his body controls every aspect. This concept is shown visually as Chris floating in a sort of space, or void, unable to control his actions yet able to see through his eyes.

Thankfully, Chris fights his captors to "get out" as the title urges. He avoids becoming like the victims before him, escaping his fate of living in the sunken place. His first experience of the sunken place is brought on as Missy Armitage, Rose's mother, utilizes her knowledge of hypnosis to discern how susceptible Chris would be to mind control. She aptly clinks a spoon against the side of a teacup, creating a hypnotic rhythm that sends Chris floating.

How effective is hypnosis, and could it really be used on unsuspecting victims? Hypnosis is defined as a:

Special psychological state with certain physiological attributes, resembling sleep only superficially and marked by a functioning of the individual at a level of awareness other than the ordinary conscious state. This state is characterized by a degree of increased receptiveness and responsiveness in which inner experiential perceptions are given as much significance as is generally given only to external reality.[2]

We have all seen depictions of hypnosis in media. Vampires, for example, use a sort of charismatic mind control to lull their victims. There are also entertainment hypnotists, like Erick Kand, who explains on his website what to expect at his shows; "the hypnotic trance state creates a sense of heightened awareness that brings out the best in the volunteer performers. Your volunteers role-play in various hypnosis comedy routines that have your audience doubled over with laughter."[3] This is the classic "cluck like a chicken" sort of hypnosis that has been performed on stage since the eighteenth century. Although Kand and his counterparts assure we will be laughing at the hilarity that hypnosis can bring, the practice of stage hypnotism has been met with healthy skepticism. Kreskin, an American magician who performed comedic hypnosis for decades eventually maligned his former work:

For nineteen years I had believed in . . . the sleeplike "hypnotic trance," practicing it constantly. Though I had nagging doubts at times, I wanted to believe in it. There was an overpowering mystique about putting someone to sleep, something that set me and all other "hypnotists" apart. We were marvelous Svengalis or Dr. Mesmers, engaged in a supernatural practice of sorts. Then it all collapsed. For me anyway.[4]

While stage hypnotists are strictly there for entertainment, hypnosis has been used for far more serious endeavors, including therapy, and to recall traumatic events from the past. Hypnotherapy is utilized in a number of different mental health scenarios. In a study published in 2014, researchers studied whether it worked for people suffering from clinical depression.

Sachin K. Dwivedi and Anuradha Kotnala from the Department of Clinical Psychology in Hardwar, UK, concluded that hypnotherapy, as long as it was done by professionals with the proper approach, *did* improve the lives of people who were afflicted with depression:

> Modern hypnotherapy is widely accepted for the treatment of anxiety, subclinical depression, certain habit disorders, to control irrational fears, as well as in the treatment of conditions such as insomnia and addiction. Hypnosis has also been used to enhance recovery from non-psychological conditions such as after surgical procedures and even with gastroenterological problems including Irritable Bowel Syndrome.[5]

Whether or not stage hypnotists are authentic, there is proven science behind the efficacy of hypnosis in a medical setting. Does this also include the recall of long buried memories? Are memories drawn from hypnosis to be believed? According to Dr. Brian Thompson, a licensed psychologist at Portland Psychotherapy Clinic, Research, and Training Center in Portland, Oregon, "scientific data suggests that the use of hypnosis in recovering memories is very problematic."[6] Thompson goes on to assert that hypnosis hurts rather than helps:

> Not only is hypnosis no better than regular recall, data suggest that recall during hypnosis can actually result in the creation of more false memories than recall while not under hypnosis. Furthermore, people who recall memories under hypnosis are more likely to believe in the accuracy of these memories, regardless of whether they are true or not. It is for these reasons that many professionals working with individuals who may have been abused as children strongly caution against the use of hypnosis as a tool to try to recover possible unremembered trauma. The American Medical Association took a stand warning against accuracy of memories recovered through hypnosis in 1985.[7]

The resounding answer about hypnotic memory recall is that it is not reliable. A research study in 1988 concurs; ". . . extreme caution should be exercised in employing information and impressions derived from hypnotic early recollections in forensic situations."[8] It is surprising, after unearthing all this data, to realize that hypnotic memory recall has been used in the conviction of crimes. One such case involves the "Dating Game Killer" Rodney Alcala. A serial killer responsible for countless, brutal deaths spanning the country, Alcala was granted a retrial in the 1979 murder of Robin Samsoe because of the prosecution's use of hypnotic memory recall. After the overwhelming scientific data compiled in the 1980s, it became evident that any witness statements derived from this practice needed to be dismissed. During the original Alcala investigation, a witness, US Forest Ranger Dana Crappa, underwent police-sanctioned hypnosis to better remember the vehicle and suspect she'd seen. Because of this, Alcala was given another day in court. Fortunately, there was overwhelming evidence against Alcala without Crappa's statements, therefore the "Dating Game Killer" did not escape justice. This reminds us of a futuristic *Black Mirror* (2011–Present) episode, "Crocodile" in which insurance companies are able to access memories in order to dispute claims. Imagine the future of criminal justice if this were possible!

Hypnosis is a proven state of being. Although, it's proven that memories drummed up from this autonomic, sleep-like trance are not dependable. But, as in the case of Chris Washington in *Get Out*, can someone be hypnotized without giving their consent? This can perhaps be answered by the bizarre case of former attorney Michael

Hypnosis is a proven state of being.

Fine, who was sentenced to twelve years in prison for multiple counts of "kidnapping with sexual motivation" in 2016. Yet, this is not a clean-cut case of a bad man tying rope around his victim. Instead, Fine was accused of hypnotizing his clients in order to have sex with them while they were in their fugue state. One unidentified victim described how she had no memory of anything untoward after her first meeting with Fine. Although when she left his office her bra was unclasped and she had a peculiar feeling. Also, her memory was blurry, as though she couldn't recall the meeting. Disconcerted after several meetings, this victim decided to secretly audio tape her meeting with Fine. Afterward, she played the tape, shocked to discover that they had had a sexual conversation that she had no memory of! And at the end of their discussion, Fine instructed the woman not to remember anything of their meeting in the tone of a stage hypnotist. This victim was not alone, as five others came forward:

> In one complaint filed against Fine, a victim described how he would perform the actual hypnosis. Under the guise of introducing her to meditation and mindfulness, he would ask her to sit in a chair, perform a few breathing exercises and close her eyes. Sometimes he would ask her to watch the space between his fingers.[9]

Fine is not the first man to use hypnosis for lascivious means. In 2012 a physician's assistant in Michigan was convicted of second-degree criminal sexual conduct after victimizing two women after hypnotizing them. While *Get Out* is a fictional story rooted in a science that is not real (yet!) the sunken place feels authentic. The vulnerable state of those in hypnosis, as shown by the women victimized by Michael Fine, is a raw and scary place where evil people can assert their control. In *Get Out*, filmmaker Jordan Peele asks us to recognize the contrasting landscape that minorities face every day, a stark difference from the "perfect" suburbia of white privilege. What better way to do this than to introduce us to the sunken place?

SECTION SIX
DEADLY ANIMALS

CHAPTER SIXTEEN

CUJO

Year of Release: 1983	
Director: Lewis Teague	
Writer: Don Carlos Dunaway, Lauren Currier	
Starring: Dee Wallace, Daniel Hugh-Kelly	
Budget: $8 million	
Box Office: $21 million	

Just the name *Cujo* conjures images of snarling teeth and bestial violence. Ever since Stephen King's novel debuted in 1981, *Cujo* has come to describe any sort of vicious dog. While the TV series *Lassie* (1954–1974), and films like *101 Dalmatians* (1961) and *Beethoven* (1992) showcase the lovable traits of human's best friend, King's novel and its film version *Cujo* (1983) depict what happens when dogs go bad. In the first scenes of the film, Cujo appears how we would normally assume a dog to be. He seems happy and curious, enjoying a summer day in the fields of Maine. Lassie's day would have ended there, with nothing more treacherous than perhaps having to save a child who'd fallen down a well. But for Cujo, this is when the horror begins. On a hunt for a spry rabbit, Cujo sticks his fluffy head into a cave. Suddenly, a bat bites poor Cujo on the nose! While these sorts of misadventures happen in any number of children's movies (remember when Milo the cat gets his lip pinched by a crab in *Milo and Otis* (1986)? *Cujo* is a horror film from the get-go. It is this bat's bite that sets in motion the tragic story of Cujo the dog and the unfortunate humans around him.

Bats have long been characterized as villains. Count Dracula and his contemporaries could transform into the winged creatures, spreading vampirism across the lands of Europe and beyond. In *The Abominable Dr. Phibes* (1971) a victim is murdered by a large, bloodthirsty bat, secreted in his room for the purpose of his demise. A bat also attacks in *Suspiria* (1977), getting caught in the long hair of main character Suzy Bannion (Jessica Harper). So, what is it about bats

Bats can be carriers of rabies but very few are.

that cause many of us to shudder? Biologist Elizabeth Hagen can guess why people have a natural aversion to the nocturnal animals. "Most people are afraid of bats because they think that all bats have rabies."[1] This, of course, is the case in *Cujo* and perhaps this story has perpetuated the myth that bats are more likely to carry rabies than other animals. Hagen insists that in reality "very few bats have rabies."

Dogs, too, can strike fear. This is less common, as dogs are often portrayed as kind, cute, and innocent in media. Cynophobia, the clinical term for the "irrational and persistent"[2] fear of canines, is often brought on by a traumatic experience with a dog. Phobias can bring about both physical and emotional symptoms. An important distinction is that a person suffering from cynophobia would be afraid of Cujo even before he was stricken with rabies. Most of us don't have cynophobia, and with over forty-two million[3] households with dogs in the United States alone, they are a ubiquitous piece of our cultural puzzle. Despite their prominence in our lives as symbols of good, there have been occasions when they were characterized as monsters long before *Cujo*. If the "boy next door" can be frightening, as in the case of gentle-seeming Norman Bates of *Psycho*, then what is more horrific than the darling creature at your feet turning on you?

Cujo is not the first dog in literature to have rabies. Meg remembers the emotional reaction she had to Tim Johnson's death in Harper Lee's masterpiece *To Kill a Mockingbird* (1960). It's a scene, both in the book and in the 1962 Gregory Peck film, that has stuck with her. Tim, a happy-go-lucky dog, is infected by rabies. He becomes "mad," foaming at the mouth and snarling as menacing as Cujo. Many a high school teacher has speculated over the symbolism of Tim Johnson. The poor dog has been described as an emblem of racism, mob rule, and injustice. He is then killed by Atticus Finch, a necessary compassion for the animal, and a preventative measure to avoid human death.

If only Atticus Finch had lived in Castle Rock.

Aside from the obvious association our beloved pets have with the werewolf, there is a supernatural being known as the "black dog" which shares even more canine traits. The folklore surrounding black dogs dates back to the sixteenth century when "on August the 14th in the year 1577 the 'Black Dog' was responsible for killing a mass of people who were praying in a Church which was situated in East Anglia."[4] From its inception, the black dog has found particular purchase in the UK, from the beaches of Cornwall to the moors of Yorkshire, where sightings of these unnatural beings have been reported. These black dogs are larger than any house pet, and, interestingly, they are considered to be nocturnal and have glowing eyes, which is a striking link to the feared bats. They are unlike werewolves in that black dogs are believed to be ghosts or demons, canine creatures that exist in the hazy border between our world and the world of the dead.

Black dogs have been explored in Sir Arthur Conan Doyle's novel *The Hound of the Baskervilles* (1902) which has been the subject of many film and television adaptations, including the 1939 version starring Basil Rathbone. In the novel, Doyle takes inspiration from the folklore rampant in England, adding in the concept of the black dog being more than an apparition. "But that cry of pain from the hound had blown all our fears to the winds. If he was vulnerable, he was mortal, and if we could wound him, we could kill him."[5] Before Doyle was even born, Charlotte Brontë made mention of the ominous black dog, also known as a "gytrash" in her gothic masterwork *Jane Eyre* (1847):

It was very near, but not yet in sight; when, in addition to the tramp, tramp, I heard a rush under the hedge, and close down by the hazel stems glided a great dog, whose black and white colour made him a distinct object against the trees. It was exactly one form of Bessie's

Gytrash—a lion-like creature with long hair and a huge head: it passed me, however, quietly enough; not staying to look up, with strange pretercanine eyes, in my face, as I half expected it would. The horse followed,—a tall steed, and on its back a rider. The man, the human being, broke the spell at once. Nothing ever rode the Gytrash: it was always alone; and goblins, to my notions, though they might tenant the dumb carcasses of beasts, could scarce covet shelter in the commonplace human form. No Gytrash was this,—only a traveler taking the short cut to Millcote.[6]

In film history, the black dog has made several appearances. There is the 1978 American made-for-television film *Devil Dog: The Hound from Hell* with an unintentionally hilarious VHS cover of a dog with devil horns growling beneath a crudely drawn pentagram. The movie starred Richard Crenna, and centered on a suburban family who accidently adopt a puppy from hell. There is also the aptly named *Black Dog* (1998) starring Patrick Swayze. *Black Dog* is more action and less horror, although it compounds on the black dog legend, developing it into a frightening vision seen by truckers on lonely stretches of highway.

Cujo the novel has one of Stephen King's most bleak endings (spoiler alert: everyone dies!) so it's no surprise that screenplay writers Don Carlos Dunaway and Lauren Currier tweaked the ending. They hoped for a more crowd-pleasing finish, with Donna Trenton (Dee Wallace) and her son Tad (Danny Pintauro) surviving the rabid dog-monster of nightmares.

As we researched black dogs, *Cujo*, and bats, we became curious about rabies. Is Cujo's behavior depicted realistically? Have rabid dogs killed humans? And is rabies a real threat in modern America or anywhere else in the world? According to the World Health Organization (WHO), rabies is a preventable disease that still kills humans in over 150 countries. Ninety-five percent of these deaths occur in Africa and Asia:

Rabies is one of the neglected tropical diseases that predominantly affects poor and vulnerable populations who live in remote rural locations. Although effective human vaccines and immunoglobulins exist for rabies, they are not readily available or accessible to those in need. Globally, rabies deaths are rarely reported and children between the ages of five to fourteen years are frequent victims. Treating a rabies exposure, where the average cost of rabies post-exposure prophylaxis (PEP) is US $40 in Africa, and US $49 in Asia, can be a catastrophic financial burden on affected families whose average daily income is around US $1–2 per person.[7]

While rabies is not a threat in the modernized world, children in third-world regions seem to be the most at risk. A shocking statistic from WHO purports that the overwhelming majority of human rabies infections come from dogs: 95 percent. This makes sense, as dogs live more closely to humans than most other species, and also lends credence to the believability of *Cujo*.

Rabies inflames the brain of the victim, whether animal or human, and once symptoms are present there is little hope of survival. It first presents as a rather general sickness, with nausea, fever, and sore throat. As the disease spreads throughout the brain, it will eventually cause seizures, paralysis, and coma. One symptom of rabies would be familiar to fans of *Cujo*. Rabies has been proven to cause aggressive and irrational behavior. This is depicted in Cujo's sudden personality change as he quickly changes from lovable pup to vicious killer.

There are rare instances of rabies killing people in the UK and the US. Omar Zouhri died in 2018 at John Radcliffe Hospital in Oxford, England, after being bitten by a rabid cat during a vacation to Morocco. When he complained to his doctor of hand paralysis, the first rabies test came back negative. Unfortunately, by the time the doctors at Radcliffe realized it was rabies, it was too late for Mr. Zouhri. Before this, the last recorded rabies death in the UK had occurred in 2012 once again after the victim came home from a vacation, this time in Asia, where he'd been bitten by a dog. A similar tragedy happened in the United States in 2017. The Centers for Disease Control outlined the death of a sixty-five-year-old woman who vacationed in India. While there she was bitten by an aggressive puppy. Six weeks later, back at home in Virginia, the woman complained of pain and tingling in her right arm. In a few days the woman had insomnia, trouble swallowing water, and intense pain. She was diagnosed with having anxiety. Alex Berezow from the *American Council of Science and Health* described the deterioration of her body:

> The patient's condition worsened, and she was taken by ambulance to a different hospital. Now, she was displaying a lack of coordination, which often indicates some sort of neurological problem. Doctors had reason to believe she was suffering from a blockage in a heart blood vessel, so she underwent an emergency catheterization (i.e., doctors stuck a tube in her heart). They found nothing abnormal. By that night, the patient was agitated, combative, and gasping for air when trying to drink water. That's when the medical team learned from her husband that she had been bitten by a dog in India.[8]

She died after many life-saving measures, including the "Milwaukee Protocol" which is an experimental treatment in which the patient is put into a coma and given ketamine, ribavirin, and amantadine. This mix of drugs were chosen to attack the inflammation in the victim's central nervous system. Since its first use in 2004, the Milwaukee Protocol has been tried twenty-six times, being successful only once. The single person saved by this treatment, a fifteen-year-old girl, left the hospital after seventy-six days and went on to attend college.

Rabies is a rare way to die in the modern age, yet it is not unheard of. The WHO is working to eradicate rabies deaths from the globe, hoping to meet this goal by 2030. The likelihood that an American dog would suffer from rabies is slim to none, but Cujo's aggression seems realistic in both the book and the film. And so, the palatable fear of rabies, of its ability to kill so swiftly and painfully, remains.

CHAPTER SEVENTEEN

ARACHNOPHOBIA

Year of Release: 1990	
Director: Frank Marshall	
Writer: Don Jakoby, Wesley Strick	
Starring: Jeff Daniels, Julian Sands	
Budget: $22 million	
Box Office: $53.2 million	

Spiders are scary. Their spindly legs, tiny fangs, and bulbous group of eyes terrify. It's quite nearly a fact, as they have long been derided in popular culture. For every *Spider-Man* there are dozens of spider-centric horror movies, think *Tarantula* (1955), *The Giant Spider Invasion* (1975), and *Eight Legged Freaks* (2002). According to Claire McKechnie in her study for the *Journal of Victorian Culture*, spiders were once observed as being associated with "ingenuity and industry"[1] but as gothic literature ascended in popularity through the late Victorian age, the spider became emblematic of the dark and mysterious. "Much maligned as the unfamiliar Other, the spider caused—and mitigated—anxieties about the limits of the human." Perhaps it is this "otherness" that is at the heart of our collective fear of spiders. In a 1991 study at the City, University of London, Graham Davey concluded that it was, indeed, their distinctly non-human differences that instill fear. "It turns out that it is not so much a fear of being bitten, but rather the seemingly erratic movements of spiders, and their 'legginess.' Davey said animal fears may represent a functionally distinct set of adaptive responses which have been selected for during the evolutionary history of the human species."[2] This makes sense, as the "erratic movements" of spiders, or other creatures, would have tipped off our ancestors to potential dangers. Whatever the reason, as we sift through research for this chapter, we feel as though spiders are scrambling up our spines!

"A return to the spoofy scares of the '50s drive-in movies,"[3] *Arachnophobia* hit theaters in July of 1990. Starring Jeff Daniels as spider-phobic Dr. Ross Jennings and John Goodman as a goofy yet determined exterminator, *Arachnophobia* was a Disney produced picture with a sizable budget. In an interview with the *Orlando Sentinel*, Daniels explained the tone they were going for. "We don't have chainsaws going through necks and blood spurting. It's scary, but this is not *The Attack of the Killer Spiders*. We approached it as a comedy with a couple of thrills. We knew we had the thrills in there, so we worked hard to make sure the movie had a sense of humor about itself." The humor, he said, "kind of relaxes the audience, so that we can come in and get them again."[4] So don't let the Disney name and humorous bent fool you, the spiders in *Arachnophobia* play upon our shared fears of spiders' "otherness." The film starts in the Amazonian rainforest where enormous dead spiders are literally falling from towering trees. After an entomologist dies from a spider bite his body is sent back to his hometown in America. This concept of dying from a spider bite, particularly in the Amazon, is not fiction. A small arachnid called the wandering spider is considered one of the deadliest animals in the Amazon. If disturbed by humans it will bite. "The venom of the spider causes extreme pain and inflammation, as well as loss of muscle control which might lead to respiratory paralysis and death."[5] Before an antidote was created in 1998, fourteen people were killed by the wandering spiders' bites. In *Arachnophobia*, one deadly spider, much larger than the real wandering spider, is mistakenly sealed in with the corpse of the entomologist, which becomes the catalyst to spiders running amok in the sleepy, idyllic town of Canaima, California.

In the film, the venomous spiders come from outside of the country, but it is vital to note that America is home to deadly spiders, too. A study of American spider bites between 1934 and 2014 concluded that two such species were to blame for nearly all spider-related human injury or death. These are *Loxosceles reclusa* (brown recluse) and *Latrodectus mactans* (black widow). "A brown recluse bite can take up to six weeks to heal, and in serious cases, patients can take months to recover from necrotic ulcers, fever and general malaise." In the case of black widows:

> Symptoms may include tender lymph nodes, muscle pain, nausea and abdominal rigidity with no tenderness. The degree of envenomation depends on several factors, including the amount of venom injected, size and species of spider, time of year, size and age of the victim and location of the bite. In patients with other underlying health issues, cardiovascular issues and even death may occur.[6]

Director Frank Marshall wrangled two real, non-American spider species to be used in the production. The spider who takes a ride in the casket from the Amazon needed to be as huge as possible, so a bird-eating tarantula was "cast" in the role. "Big Bob," named after fellow director Robert Zemeckis, had a chest prosthetic attached as well as purple stripes painted to his back in order to appear even more exotic and menacing. The plot calls for this hitchhiking spider to mate with a domestic spider, which meant *Arachnophobia* needed an army of spiders to terrify both Dr. Jennings and the moviegoers. After a series of tests, including speed trials and a sort of "spider olympics" Marshall and his team chose three hundred Delena spiders to make their feature film debut. Delena spiders, or "Delena cancerides, the flat huntsman spider or social huntsman spider, is a large, brown huntsman spider native to Australia. It has been introduced to New Zealand, where it is sometimes known as the Avondale spider as they are commonly found in the suburb of Avondale, Auckland."[7] These thankfully harmless spiders were chosen because of their size, quick movements, and sinister look. Spiders of the same species were also employed by director Sam Raimi for the 2002 *Spider-Man* film starring Tobey Maguire.

How can a director possibly control three hundred spiders on a film shoot? It became clear that Marshall, or any Hollywood director, hires experts for such a monumental task. In the case of *Arachnophobia*, bug-wrangler Steven Kutcher was brought in to make certain filming went smoothly. In an interview with *Entertainment Weekly*, Kutcher shared several of his clever techniques to ensure that the temperamental eight-legged actors took their cues:

> Arachnids are phobic about Lemon Pledge (it gums up their feet). Kutcher was able to control some of their movements by spraying blotches of the stuff on the sets' floors and walls. Also, spiders don't like heat, so hair dryers blown through pinholes were effective prods. For more exact choreography, minuscule leashes were attached to their abdomens with wax. And in some extreme instances, tiny metal plates, controlled by electromagnets, were glued to their tummies. A technique, Kutcher assures, which causes them no harm.[8]

Although real spiders were used in the majority of the shots, rubber spiders had to be used when the script called for a spider's death (no spiders were killed in the making of *Arachnophobia*). During the climax when Ross Jennings must defeat the vicious "Big Bob," a mechanical fifteen-inch double had to be made. "He has to stalk Jeff Daniels; he has to stay in the right light, and if we waited for him to do that, we'd be here three or four months longer," Marshall told the *New York Times*. "The main character had to become a creature, and no spider out there could give us the vicious, evil close-ups the script called for," added visual effects supervisor David Sosalla. "The evilest ones, with real ugly looking faces, were too tiny."[9] This suggestion that the spider needs to become a creature and more, display humanistic traits of evil, is important to note. Is attributing negative human characteristics to animals what ultimately makes for an effective movie monster?

The tendency for us to typify animals as having similarly human emotions is one rooted in our childhoods. "Children are frequently exposed to anthropomorphic depictions of animals. The impact of anthropomorphism on children's development of factual and biological knowledge about real animals has consequences for how we engage children in early learning about the natural world."[10] This is often in a positive light (especially in our view of household pets) and even in the depiction of the pleasant spider Charlotte in E. B. White's novel *Charlotte's Web* (1952). Charlotte is the heroine of the story, which is later adapted in animation (1973) and live action (2006) films. She is kind, patient, and saves Wilbur from slaughter. And her death is one of the more poignant deaths in children's literature. If Charlotte's cleverness and helpfulness can transform a simple barn spider into a heroine, then it stands to reason that the makers of *Arachnophobia*, and so many killer-spider films of its ilk, must portray spiders as having malevolent intentions. The spiders' "legginess" might be what makes us cringe, but it is their sinister machinations of death and destruction that make a movie. Cujo, too, is anthropomorphized. Rather than characterized as an animal who is confused, unlucky, or a victim, he is seen as innocent. And at the turn of a bite, he is at once evil.

An important aspect of *Arachnophobia* is revealed in the title itself. Main character Dr. Ross Jennings is afflicted with arachnophobia. A fact known to his wife and children, as his wife, Molly (Harley Jane Kozak), usurps gender norms by being the spouse who dispatches a spider in their home. Ross explains that this phobia began in childhood when he was terrorized by a spider in his crib and was unable to move out of fear. This paralysis returns at the climax of the film when Ross is struck motionless as "Big Bob" walks on his body. In true movie fashion, Ross overcomes the paralysis and ultimately kills the creature and restores his family and town back to spider-less normalcy. This film, and the

Arachnophobia is the fear of spiders.

trope of deadly animals in general, got us thinking about phobias. How do they manifest physically? And is Ross Jennings's temporary paralysis a realistic depiction of a phobic's response?

More than twenty-six million people in the world suffer from a phobia, an exaggerated or irrational fear which can be classified into one of three categories; simple phobias, social phobias, and panic attacks. Arachnophobia, along with intense fears of things like snakes, closed places, and even clowns falls under the simple phobia classification. As for the second category:

Social phobias are fears of being in situations where your activities can be watched and judged by others. The difference between having a social phobia and simply being shy is that shy people usually don't try to avoid social situations. People with social phobias find excuses not to go to parties or out on dates. If asked to give a speech in class, they react as if they are facing a real physical threat.[11]

Panic attacks, the last form of phobia, greatly alter a person's life. An example would be an agoraphobic, who while out in a public place suddenly feels physical symptoms like lightheadedness. They feel they are in great danger, and thus choose to stay out of public places, further intensifying their phobia. In the case of Ross Jennings in *Arachnophobia*, it seems appropriate to say he is indeed in this first "simple" phobia class, as his life is not vastly altered by his fear of a specific trigger, and the story of Ross's childhood trauma rings true under the scientific studies of phobias. A moment or series of moments in our youth can construct a fear in our minds that takes hold. "Although childhood fears are a part of normal development, a significant minority of children evince fears that interfere with their functioning (specific phobia in the DSM-IV)."[12] For those whose fears do not diminish over time, these specific or simple phobias become a part of their adult life.

Ross Jennings has a physical reaction to his specific phobia. He loses motor function, describing a sensation of not being able to move, no matter how badly he wants to. We dug into medical literature to find if this is an authentic representation of a person suffering from a phobia. The Mayo Clinic describes several anxiety-based physical reactions for coming face-to-face with whatever specific phobia encountered, including nausea, dizziness, fainting, profuse sweating, rapid heartbeat, and a tight chest. Temporary paralysis is not listed. We broadened our search to find out if anxiety, in whatever form, can spark temporary paralysis. According to clinical psychologist Daniel Sher, MA, this is indeed the case. Sher gives two reasons for why this loss of motor function could occur under anxiety inducing conditions. The first is hyperventilation, depicted in many films as short breaths only remedied by exhaling and inhaling into a brown paper bag. Hyperventilation is actually the act of breathing out too much carbon dioxide. It is triggered by anxiety, which the Mayo Clinic describes as the main emotion felt by someone faced with a phobia. Hyperventilation can cause a person's limbs to go numb, making them feel as though their muscles are not able to move. Secondly, Sher reasons why someone under great distress would sense that they are unable to control their body:

> When someone suffers from anxiety, they often focus deeply on the way their body feels, becoming highly attuned and conscious of movements which would otherwise be performed spontaneously and automatically. The process of actively contemplating the series of movements that you're performing may interfere with the automatic process whereby those actions would normally be carried out. This may make automatic movements harder to perform, creating a sense of immobilization.[13]

Sher further contends that while we discuss the concept of the "fight or flight" response to fears and anxieties, we should tweak it to be "fight, flight, or freeze." The concept of a deer in headlights is proven to afflict humans as we, too, can freeze at the sight (or, in the case of the film, touch) of our greatest nightmares. We've concluded that because anxiety greatly affects all functions of a person's body, including hyperventilation, Ross Jennings's physical response is indeed an accurate portrayal of a phobia in the aptly titled *Arachnophobia*.

CHAPTER EIGHTEEN
THE BIRDS

Year of Release: 1963	
Director: Alfred Hitchcock	
Writer: Evan Hunter	
Starring: Rod Taylor, Tippi Hedren	
Budget: $3.3 million	
Box Office: $11.4 million	

If dogs are a ubiquitous piece of our cultural landscape, then we are positively drowning in birds. Estimating how many birds there are on Earth proves to be difficult, but a conservative guess would be about one hundred billion.[1] (Meg gasps in horror!) Birds perch on our mailboxes, eat seed from our feeders, and scrounge for food in our picnic baskets. They are everywhere, whether we live in the bustling city center of London or on a rural farm in the Midwest of America. Birds are often seen as little more than an aspect of the background; they exist on the roads, up in telephone poles, swarming trees, even as pets in gilded cages. It is this prevalence that creates the unsettling question of what would happen if birds became malicious. Unlike spiders, snakes, or fellow-winged bats, birds are seen by nearly every human every single day. One is to assume that this omnipresence creates a sense of safety, that we consider birds to be harmless as they are constantly within view. We even delight in spotting different species. Yet, we have to wonder, are we watching the birds, or are they watching us?

Daphne Du Maurier, author of the well-known gothic novel *Rebecca* (1938), asked this very question in her novella *The Birds* published in her story collection *The Apple Tree* (1952). In *The Birds,* a small town in Cornwall, England is suddenly attacked by seabirds. The attacks are brutal with no answer as to why. Unlike *Arachnophobia* in which the catalyst of the film is an Amazonian spider coming to the United States, and in *Cujo* in which a dog is bitten by a rabid bat, the bloodthirsty birds of *The Birds* have no specific trigger or reason for their brutality. Perhaps it is this randomness that creates a whole new layer of horror to the trope of deadly animals.

Alfred Hitchcock seemed to think so. In 1940 he had adapted Du Maurier's *Rebecca* into a successful picture starring Laurence Olivier and Joan Fontaine, so it was no wonder he dove back into Du Maurier's source material, choosing to base another one of his suspenseful films on her novella. Before hiring screenwriter Evan Hunter to adapt and tweak the plot of *The Birds,* Hitchcock did his own research on a real-life bird "attack" in California. In 1961, a few mere weeks after this incident, the director requested news copy from the *Santa Cruz Sentinel* to study.

He read of the eerie morning of August 18th, 1961, when residents of Capitola, California, a smaller community on the coast of Monterey Bay, awoke to a frightening discovery. Droves of sooty shearwaters, a medium sized seabird native to the area, were acting erratically. Some crashed into rooftops, windows, and cars. Others flopped, dying in the streets, while more vomited their fish dinners into the grass. It was a short incident, but one that terrified and traumatized those who had witnessed this mass bird hysteria. Desperate to find an answer for this sudden behavior change of the seabirds, scientists hypothesized that domoic acid poisoning could be to blame. This was never proven, but it is a viable suggestion. Domoic acid is a neurotoxin produced by algae. These algae can then accumulate in shellfish, sardines, and anchovies. When animals or humans ingest this toxin, which was ingested by

Domoic acid structural formula.

the shellfish, their brain is affected, causing seizures and even death. "Domoic acid is a tricarboxylic amino acid that is classified as an excitatory amino acid (along with the dicarboxylic amino acids glutamic acid and aspartic acid). It acts through the inotropic non-NMDA receptor and especially affects the hippocampus and amygdala of the brain. The time from ingestion to intoxication can range from minutes to hours."[2] Seabirds would undoubtedly be susceptible to this type of poisoning, and the description of the event in August of 1961 mirrors the listed animal symptoms of domoic acid poisoning: head weaving, seizures, bulging eyes, mucus from the mouth, disorientation, and death. In 1991, domoic acid was the official cause of death for hundreds of brown pelicans and cormorants in Monterey Bay, on the same beaches the peculiar bird activity had played out exactly thirty years earlier. A bird attack in 2006 was also attributed to domoic acid poisoning when a brown pelican burst through the windshield of a moving car on the Pacific Coast Highway.

Unfortunately, humans have also been victim to this poisoning. The most dramatic case occurred in 1987. Over one hundred people in Eastern Canada fell ill after ingesting mussels fished from the coast of Prince Edward Island. They described a number of symptoms including nausea, vomiting, severe headache, and most alarming, loss of memory. This amnesia, for some, was permanent. At the time, the doctors called this phenomenon "amnesic shellfish poisoning." Three of the sufferers died from this painful syndrome caused by domoic acid. In 1991, two months after the bird deaths in California, two dozen people were struck ill with amnesic shellfish poisoning, or domoic acid poisoning, in Washington State. They had all consumed razor clams seized from the coasts of Oregon and Washington. When testing was done after the incident, it was also found that Dungeness crabs in the area were rife with domoic acid. Though human incidents of amnesic shellfish poisoning are rare, according to the Channel Islands Marine and Wildlife Institute, animal incidents seem to be on the rise.

It is generally accepted that the incidence of problems associated with toxic algae is increasing. Possible reasons to explain this increase include natural mechanisms of species dispersal (currents and tides) to a host of human-related phenomena such as nutrient enrichment (agricultural runoff), climate shifts or transport of algae species via ship ballast water.[3]

Birds are not only susceptible to domoic acid; they have also been documented as acting erratically while "drunk." In the fall of 2018, Gilbert, a tiny town in Minnesota (which also happens to be the hometown of Meg's in-laws), was thrust into the national spotlight thanks to a drove of drunken birds. Residents began to complain about robins and other small birds crashing through windows, falling out of trees, and hitting cars. While not as dramatic as the seabirds on the coast, these small species caused confusion in the Minnesotan community. There is some dissension on the cause. Gilbert Police Chief Ty Techar explained that some had got a little more "tipsy than normal." This can be caused by fermented berries, which due to the early frost in Minnesota, may have become more potent. Kenn Kaufman, field editor of *Audubon*, said "drunken birds were certainly a real phenomenon. Apart from berry eaters, drunkenness can also befall yellow-bellied sapsuckers that feed on fermented tree sap."[4] But other experts reject the notion that the birds in Gilbert were inebriated, suggesting that the robins were frantically attempting to leave town at the sight of hawks, who were migrating at that time. Whatever the cause of the strange behavior, it is safe to conclude that birds can become more than just innocuous background noise in sudden, brutal encounters.

Although he'd read the scientists' hypothesis of domoic acid poisoning in the 1961 Capitola incident, Alfred Hitchcock chose not to include any sort of scientific explanation in *The Birds*. The master of suspense knew that the mystery of the unknown was much scarier than any seaborne toxin. Also, in reality, the birds off the coast of Monterey Bay hadn't purposely attacked humans. This, once more, prompts the notion of anthropomorphism in the development of animal movie monsters. The birds in *The Birds* had to have a maliciousness in order to terrify. They are depicted as waiting patiently before an orchestrated attack on schoolteacher Annie Hayworth (Suzanne Pleshette) and her students. This happens in great effect during the diner scene, too, as the birds work to cause chaos. Which led us to question the intelligence of birds and their ability to work together, whether for productive or nefarious means.

Generalizing all birds would be a naive practice, as they range greatly in size, lifestyle, and therefore intelligence. We will focus on crows, as their mental acumen is of great scientific interest. In recent years, corvids (crows, ravens, rooks, and jays) have been praised for their rather impressive smarts. (They are also a pervading symbol of horror, thanks not only to Hitchcock, but also to Edgar Allan Poe's haunting 1845 poem "The Raven.") Studies have proven that they utilize tools, problem solve, and can consider future outcomes. Crows in Japan figured out that carefully placing nuts on the street led to cars crushing the hard shells. They were then observed making note of traffic lights before retrieving the opened nuts while vehicles idled.

Scientists contend that corvids are as intelligent as apes. They exhibit traits that were long believed to be only attributed to primates (as well as dogs and dolphins). These include recognizing themselves in a mirror, reasoning out complex problems, and using and understanding a symbol system. Researchers from the University of Iowa along with contemporaries at Lomonosov Moscow State University proved that crows could understand symbols by their ability to match like-pairs:

> To reach that conclusion, the scientists trained crows to recognize whether two objects were identical or different, which the birds indicated by pressing one button when shown pictures of objects that matched and a different button when the objects didn't match. Once all the birds were good at matching objects, researchers showed the crows images of pairs of objects. Some images depicted matched pairs, while others depicted two mismatched objects with different shapes or colors. In response, crows could press buttons to choose between a matched pair or a mismatched pair.[5]

Fascinatingly, corvids also have a measure of social intelligence that comes to us humans in later childhood. It is termed "theory of mind," the concept of recognizing that others have similar but different thoughts, and then applying this knowledge to change one's own behavior. An example would be when ravens believe a human has spotted them in the act of hiding food. They will alter their hiding spot as a result, choosing to obscure it better, as they can predict that another entity would want to steal their goodies. Crows were chosen to be the birds perched on the playground equipment in The Birds. A crackling sea of black wings, they can elicit fear in anyone, not just ornithophobes. So, do they use this power of intellect to hurt humans? Why would they?

According to Jim O'Leary, creator of the website CrowTrax, the answer to the first question is a resounding yes. His site is dedicated to tracking the data of crow attacks on a virtual map. Since CrowTrax's inception, O'Leary has documented nearly five thousand crow-human scuffles, all started by the flapping corvids. Users of the site often add accounts of their brutal attacks. They describe birds clawing at their heads, pulling out their earrings, and even waiting on their car windshield to pounce on them: "It's traumatic because it happens unexpectedly and from above," said O'Leary in an interview with *City Lab*. "I frequently get reports where they break the skin. They do draw blood at times. I've had other reports with women where the talon gets stuck in their hair. That'd be kind of a terrifying thing, if a crow attacks you and then the crow can't get out of your hair."[6] Hitchcock, it seems, was not far off in his portrayal of crows flapping at their victims' heads and waiting outside the school in an effort to maim. One fundamental difference to note is that at CrowTrax it appears these run-ins only involve a single bird.

This leads us to the why. The data at CrowTrax is quite telling, as crows attack most significantly in the spring when they have babies to protect. When a human or unlucky pet is ambushed by a crow, there's a good chance a nest is nearby. Perhaps because crows possess the aforementioned "theory of mind," they become particularly aggressive as they can predict a bleak future of injured or killed birdlings. And more chilling, corvids have the ability to actually memorize a face. If they believe a human has wronged them, they will nurse a grudge against that particular victim. In their book *Gifts of the Crow: How Perception, Emotion, and Thought Allow Smart Birds to Behave Like Humans* (2013), John M. Marzluff and Tony Angell explain an experiment professors and students at Washington State University conducted in order to prove this intellectual bird feat. They wore exaggerated "caveman" style masks when they captured seven crows and banded them, releasing them after a few minutes. A few days later, the professor wore a different mask to record the crows' reaction. They didn't appear to react. A few more days passed, and the professor returned to campus in the original caveman mask. This time, banded birds recognized him. One bird demonstrated an aggressive stance, calling a warning to fellow birds. These fellow birds, though they had never been bothered by the professor, listened to their friend, cawing at the "bad man." Three of the banded birds ultimately showed aggression to who they remembered had captured them, although it had been brief, and roughly a week before. All of this data suggests to us that crows, wholly intelligent, are terrifying in their capacity to recognize, understand, and coordinate. Unlike in *Arachnophobia*, there is no need to anthropomorphize the villains of *The Birds*, as there is already a cleverness in their beady, avian eyes that sends shivers down our spines.

SECTION SEVEN
GHOSTS

CHAPTER NINETEEN
POLTERGEIST

Year of Release: 1982	
Director: Tobe Hooper	
Writer: Steven Spielberg	
Starring: JoBeth Williams, Craig T. Nelson	
Budget: $10.7 million	
Box Office: $121.7 million	

W hen we were kids many a late night was spent imagining the potential horrors that could possibly plague us. We both loved horror movies and the feeling of being scared. Fueling that fire was the movie *Poltergeist* (1982). Seeing a realistic portrayal of a family on screen helped transport us to the world and imagine ourselves in Carol Anne's position. A creepy clown doll coming to life? Hearing voices through the television? A scary tree pulling us out of our beds? These all seemed like real possibilities after seeing this movie!

The term *poltergeist* is German for "noisy ghost." These ghosts are thought to be responsible for physical disturbances such as loud noises, objects being moved, and even some physical attacks. They haunt particular people instead of places. Some claims of poltergeist activity have been explained by psychological factors including illusions, memory lapses, and wishful thinking. We interviewed Jenny Melton and Blaine Duncan, investigators and researchers for the Twin Cities Paranormal Society, about modern ghost detecting techniques.

Kelly: **"How did you get into paranormal investigating?"**

Jenny Melton: "When I was nine, I lived in a home right outside of D.C., and that is where my paranormal experiences began. I would see shadow figures, hear the organ start playing that we had in the room next to me, the rocking chair would rock on its own, the TV in my room would turn on by itself. And this was when we didn't have remotes and had to use a knob to manually turn the TV on/off and adjust the volume!"

Kelly: **"I remember those days!"**

Jenny Melton: "I would see a man at the top of my stairs, and the overall energy of the house was pretty terrifying for me. No one else in my family experienced any of this and just thought I was crazy, but my friend who lived next door would describe all the same things happening in her house. She even described the same man that I would see."

Kelly: **"Okay . . . that's really scary."**

Jenny Melton: "My dad was in the Navy so we moved often and I remember being *so* happy when we moved out of that house. Throughout the next couple of years, I had a couple of friends pass away who also came to me in spirit, but it wasn't nearly as scary as that house in D.C. because I felt like I knew who it was that was haunting me."

Meg: **"I could see that. So, the thought of ghosts doesn't scare you anymore?"**

Jenny Melton: "After watching a few episodes of *Ghost Hunters* and being introduced to a scientific approach to looking at the paranormal, I decided to write the Fairfax County Historical Society and inquire about my old house in D.C. Fortunately, they were more than willing to do some research for me and ended up sending me a huge manila envelope with a bunch of documents and deeds regarding the house and the property it sat on. Being right outside of D.C. there was a lot of trauma tied to the land in terms of battles and death. I'm flipping through the packet they had sent me, lost in deep thought and concentration, and there is a picture of the man I used to see at the top of our stairs."

Kelly: **"Whoa. That's incredible!"**

Jenny Melton: "In that very instant, an instant I thought would have completely freaked me out, I was instead completely validated. What I had experienced was real. I wasn't crazy. With my fear completely eliminated, that day in 2004 my whole life changed as I made it a life purpose to delve more into the things I don't understand."

Meg: **"Can you describe the equipment you use in ghost detecting?"**

Jenny Melton: "So, we use a lot of different devices that detect different forms of energy. For years my only piece of equipment was a digital voice recorder. These devices can pick up a higher range of sound frequencies than are audible to the human ear. What we are looking for from a digital voice recorder are EVPs: electronic voice phenomenon. In my opinion, this is one of the least foolproof pieces of equipment we use, as long as you are

Ghost hunters use a variety of equipment for detection.

completely conscious on 'tagging' any outside noises. Something so small like a stomach growling, a whisper, or an animal making sounds outside can be picked up very easily on these recorders, so we have to make sure we mark any noise we hear so we don't mistake it as something paranormal. What we look for is a voice that is not our own, or a response to a question that we ask."

Kelly: "I've seen that in horror movies!"

Jenny Melton: "Other devices that we use measure the electromagnetic frequency [EMF] through items like a KII meter, a MEL meter, or a REM Pod. All of these items will detect any changes in EMF, but they use different techniques to relay that information to us."

Meg: "I had no idea there were so many different types of equipment!"

Kelly: "What about video equipment?"

Jenny Melton: "We also always set up four to eight infrared cameras and a DVR system anytime we do an investigation. We typically place the cameras in 'hot spots' where some of the claims have taken place. The infrared allows us to see what is going on even when it is dark. Some of us also use handheld infrared camcorders which are helpful to see what is going on around us as we sit in the dark."

Kelly: "That would creep me out!"

Meg: "Can you tell us about the differences in what you are detecting?"

Jenny Melton: "There are really only four main categories of ghosts or spirits. First is a residual energy. This is an imprint of energy that has been left in an environment that literally replays itself over and over like a movie. We often will see this with more traumatic or tragic circumstances. When you have claims of 'spirits' walking through walls, for example, this is likely a residual haunting because at one point there may not have been a wall there. I would say battlefields like Gettysburg likely hold a lot of residual energy. This isn't energy that can interact with you because it is just an imprint of energy that repeats itself over and over. An intelligent haunting is the one we most often encounter."

Meg: **"What's an intelligent haunting?"**

Jenny Melton: "This is when a ghost or spirit is well aware that you are there and trying to communicate with it. When we get responses on our equipment, such as an EVP responding to a question we asked, this is likely the spirit of a deceased human being. Demonic hauntings are the third type of haunting, and are actually much rarer than TV or movies make them out to be. In the fourteen years I have investigated, I have only come across one case that I deemed demonic. Demonic hauntings are way past our area of expertise, so in that instance I called in a Reverend trained in exorcisms to come out and do a deliverance for our client. Again, very, *very* rare, and I have never heard of a case that occurred without the 'demon' being called in. They don't just show up."

Kelly: **"That's comforting to know! So, what about poltergeists?"**

Jenny Melton: "There is a bit of controversy over what an actual poltergeist is. It could just be an intelligent spirit who is able to muster up enough energy to be able to physically move things. The second theory of a poltergeist is based on the energy of an actual living person. We often see this kind of haunting in pre-pubescent teens and even pregnant women, which has led to the theory that with the influx of hormones, the individual themselves is able to unknowingly project enough energy to produce telekinetic results.

Kelly: **"What are your thoughts on the accuracy of how things are portrayed in the movie *Poltergeist*?"**

Blaine Duncan: "For a film set in 1982, the overall accuracy of the equipment and setup is not that far-fetched from what we, as investigators, use today. I would say the process of the paranormal investigators in the film is somewhat similar to how we set up for a paranormal investigation, with cameras, handheld equipment, and other devices said to capture proof of the paranormal. Psychics and paranormal investigators go hand-in-hand in many cases, and often-times they're able to verify what each other may be experiencing, much like what we see in *Poltergeist*."

Kelly: **"Are there any horror movies or TV shows that stand out to you as getting things right in their portrayal? Or any that you think get things wrong (in regard to ghost detection)?"**

Blaine Duncan: "Overall, I don't know that anyone has come close to capturing the reality of paranormal investigating in the fictional realm of film and television. Quite honestly, if people knew how often we sit in the dark for hours and hours on end talking to nothing, they would probably find very little interest in what we do. The sexiness of paranormal investigating that you see on the big screen is far and beyond what it's like in real life. Never have I seen someone's eyes turn red or levitate like we've seen time and time again on screen, but I can tell you from personal experiences that getting grabbed by someone who isn't there or hearing a disembodied voice echo in the halls of a reportedly haunted location is much more frightening than what you see on the screen. If there was a television show that has ever done a decent job of portraying what paranormal investigating is actually like, I would say the reality television series *Ghost Hunters* did just that. There's no doubt *Ghost Hunters* spawned an interest in the paranormal for hundreds of thousands if not millions of people around the world."

Meg: **"Have you found proof of paranormal activity or been able to refute a claim?"**

Jenny Melton: "There have been plenty of cases that, through our equipment, we have been able to deem what seemed like paranormal phenomena to have completely natural explanations. The funniest thing about the paranormal field, though, is that I would be skeptical of anyone who actually said they had, without a doubt, evidence of a ghost or spirit. I have had a lot of paranormal experiences, but by definition, paranormal only means that it is out of the realm of what we as humans consider to be normal. I have had many things happen, and I have documented a lot of things that I have not been able to explain. Paranormal: yes. Is it proof of a 'ghost'? I can't necessarily claim that, and I would be wary of anyone who says that they can prove anything. I personally believe in spirits and ghosts, but it is not my place to convince anyone of anything. I can't 'prove' anything."

Kelly: **"Is there anything else you'd like people to know about what you do?"**

Jenny Melton: "There is a ton of work required, and not all of it is fun. Our process for one case could take months. We begin with a phone interview when the team in contacted. We have to make sure we are dealing with legitimate claims from people who are truly seeking our help. We then travel out to the property for a preliminary interview, which gives us a chance to get much firmer details of the claims while also making sure the property and client are safe for our investigators to come in and conduct an investigation. We have a five-page questionnaire that gets pretty detailed, but gives us enough information on the clients and the property to make a wise decision as to what our next move is going to be. If we decide to go ahead with an investigation, we then spend some time visiting historical societies and libraries looking for any information on the property that could be useful for us. We then conduct the investigation, spend about two weeks going over our video and audio, then return to the property with an extensively detailed report about exactly what happened during our investigation and what our conclusions are based on that investigation. We then make recommendations, which could involve anything from offering a space clearing of the property to 'hire an electrician,' and ultimately let the family or client choose what to do next. If they choose to do a space clearing, we will return to the property once again with our spiritual advisor, Brady, and he will conduct the clearing. By the end of the whole process, we usually have gone out to the property five or more times. Lastly, paranormal investigating is not something you can earn a living off of. All of our services are free of charge, as we are a non-profit organization. Beware of anyone who asks for compensation for any investigative purposes. The people who are truly passionate about this field and who truly want to help others experiencing unexplainable activity would never charge anything for their services."

While films like *Poltergeist* give us a fictional idea of what it might be like to pursue ghost activity, speaking with real-life investigators truly broadened our paranormal horizons. It provided insight into the reality of the work, reminding us that the dramatic sequences we see on film may be thrilling to watch, but there are real-life counterparts of Tangina Barrons (Zelda Rubinstein) in our backyard searching for the truth. Most of us would run at the first squeak of a ghost, or Carol Anne's famous utterance, "They're heeeeere!" If poltergeists ever make a racket in our homes, we'd be glad to have Jenny Melton and Blaine Duncan to call on.

CHAPTER TWENTY

THE SHINING

Year of Release: 1980	
Director: Stanley Kubrick	
Writer: Stanley Kubrick, Diane Johnson	
Starring: Jack Nicholson, Shelley Duvall	
Budget: $19 million	
Box Office: $44.4 million	

The iconic and terrifying imagery of *The Shining* (1980) continues to seep into our cultural experience. "Heeeere's Johnny!" There is manic and wide-eyed Jack Torrance (Jack Nicholson), ax in hand, fitting his face through the broken door. There are the two dead girls dressed in proper blue dresses (Louise and Lisa Burns) harkening for us to come "play forever and ever." There is the elevator opening slowly to reveal a deluge of blood. These beautifully macabre shots did not come without sacrifice. Director Stanley Kubrick was methodical, known to be demanding and difficult to work with. Just the principal photography on the film took over a year to complete, and rehearsals stretched on. Nicholson endured six weeks of practice on the bar scene alone, and Shelley Duvall, who portrayed Wendy Torrance, was said to have been so stressed by Kubrick and the long days of work that her hair began to fall out.

This hard work ultimately paid off, as *The Shining* became a critical darling. While its legacy continues as one of the most recognizable and well-loved horror films of modern cinema, it is well-known that Stephen King was less than impressed by Stanley Kubrick's take on his 1977 novel. Kubrick took creative license, altering quite a few aspects of the source material. This includes smaller tweaks, like changing the weapon from a croquet mallet to an ax, and larger overhauls, like focusing less on Jack's alcoholism and more on his inherent creepiness. While comparing and contrasting the film and the novel could fill an entire a book, we agree that both King and Kubrick created epic stories of ghosts, isolation, and madness. Or, perhaps, *The Shining* is simply about a man with writer's block? We can relate, because all work and no play make Kelly and Meg dull girls, too.

Ghosts haunt the Overlook Hotel in *The Shining* but what about the actual hotel the novel was based on? The Stanley Hotel, where Stephen King initially got inspired to write *The Shining*, is reportedly haunted. It's in Estes Park, Colorado, and was built in 1909. When King and his wife arrived at the hotel, it was closing

The Stanley Hotel in Estes Park, Colorado.

down for the season and they were the only overnight guests staying there. They ate dinner in an empty dining room while prerecorded orchestra music played. They stayed on the spacious, and eerily empty, second floor. King woke up that night from a terrifying dream about his three-year-old son being chased through the corridors of the hotel. The combination of the real-life setting and the nightmare inspired him to write the now famous book. Room 217 is thought to be haunted by Elizabeth Wilson, the hotel's head housekeeper. During a storm in 1911, she was injured during an explosion as she was lighting the lanterns in the room. She survived, though she broke both of her ankles. Guests report seeing the original owner and his wife on the staircase, a piano playing itself, and lights flickering on and off.

The character of Danny Torrance (Danny Lloyd) is able to communicate with Dick Hallorann (Scatman Crothers) through his mind, but does telepathy, the supposed communication of thoughts or ideas by means other than the known senses, exist? The term was coined in 1882 by Frederic W. H. Myers who founded the Society for Psychical Research.

Its purpose was "to approach these varied problems without prejudice or prepossession of any kind, and in the same spirit of exact and unimpassioned enquiry which has enabled science to solve so many problems, once not less obscure nor less hotly debated." Although scientific studies have been conducted since the organization's inception, no definitive proof exists of telepathic powers.

Wendy discovers the pages Jack has been typing and they are filled with variations of "All work and no play makes Jack a dull boy." Is it possible to write something without your own knowledge? Alien hand syndrome is a condition in which someone's hand may appear to have a mind of its own. Explored in another Stanley Kubrick film, *Dr. Strangelove* (1964), alien hand syndrome can affect other limbs with the affected person having no control over their actions. The medical explanation for this phenomenon is when a disconnection occurs between different parts of the brain that are engaged in different aspects of the control of bodily movement. There is no cure for this condition but studies show that by keeping the alien hand busy with a task it can be less distracting for the person afflicted. Alien hand syndrome, or something similar, has been used in several horror movies including *Idle Hands* (1999) and *Lights Out* (2016) to illustrate when a character is not aware of what their hand or hands are doing.

In *The Shining* we learn that the previous caretaker of the Overlook Hotel suffered from cabin fever and murdered his family. Does cabin fever actually exist? And if so, can it cause the sufferer to lash out violently? We know that Delbert Grady (Philip Stone), the former caretaker, also had ghosts to contend with, but would cabin fever alone cause a person to murder? Cabin fever is a colloquial term used to describe the feeling associated with claustrophobia and boredom that one feels when confined. When individuals are stuck indoors for an extended period of time they may tend to sleep, to have a distrust of anyone they are with, or to have an urge to go outside even in bad weather. Clinical psychologist Josh Kaplow, PhD, describes cabin fever as "your mind's way of telling you that the environment you are in is less than optimal for normal functioning. It's when you're in a space of restricted freedom for a period of time that you can no longer tolerate."[1] A similar phenomenon coined "prairie madness" was documented in personal letters, journals, and historical writings throughout the nineteenth century. While not an official diagnosis, prairie madness was wholly real. It afflicted European settlers used to urban environments who moved to the desolate prairies of the United States and Canada. When faced with harsh conditions and inevitable isolation, the afflicted would be struck with mentally destructive symptoms such as depression, changes in character, and even suicide. In women it manifested in crying fits, while in the book *Men, Women, and Madness: Pioneer Plains Literature* author Barbara Howard Meldrum explains that men were prone to demonstrate their depression through violence. Written long before *The Shining,* Willa Cather's novel *O Pioneers!* (1913) depicts a fictional account of such violence. After moving to the wide-open plains, main character Frank Shabata spirals into a dark depression due to prairie madness that ends in the murder of his wife and her lover.

In 1984, researchers at the University of Minnesota conducted a study on the affliction known as prairie madness, more commonly known as cabin fever. Minnesota is naturally the best place for such research, as cabin fever is not a foreign concept to those who have endured its long winters. In the study, thirty-five Minnesotans from varying communities were interviewed about their definition of cabin fever, as well as its effects on them. None of the respondents linked violent acts with cabin fever, but there are references to tension, as one man admitted; "temper gets short. Very short." And a woman described her husband in the throes of cabin fever; "He's just terrible at the end of winter. Just before spring he's a bear to be around."[2] While the character of Jack Torrance becomes more than just "a bear to be around," by the climax of *The Shining* we have found no research to suggest that cabin fever has been used as an official defense in a murder trial. Although, in 1984, coincidentally the same year as the U of M study, a base commander set fire to Almirante Brown Research Center in Antarctica at the thought of enduring yet another winter on our most isolated continent. Thankfully, he and his Argentine crew were not harmed as they were saved by an American ship. More recently, a Russian scientist in Antarctica had the dubious honor of becoming the first person to attempt murder on the vast, frozen continent. Enraged at a fellow scientist who continually spoiled the endings of books, Sergey

Savitsky "snapped" plunging a knife into the chest of the serial spoiler, Oleg Beloguzov. (Let this be a violent reminder to keep book endings to yourself!) Beloguzov's heart was nicked by the knife, but after being airlifted to the closest hospital, he mercifully lived. The men had spent the previous four years together working in isolation. According to the *New York Post*, "officials said that while the reading dispute was the final straw, the close confinement in the camp on remote Antarctica played a role in fueling the attack."[3] It seems that under the right conditions, cabin fever *can* play a role in violent attacks, particularly in extreme circumstances like Antarctic research stations. Because of these sorts of incidents, astronauts must prepare for the mental strife of isolating themselves from the entire planet. To learn more about space and cabin fever we talked to Olivia Koski, author of *Vacation Guide to the Solar System: Science for the Savvy Space Traveler!* (2017):

Kelly: **"How do astronauts deal with the extreme isolation of space?"**

Olivia Koski: "I believe that astronauts check in with doctors regularly. Or maybe I just saw that in a movie. I don't think so, though. I think they have checkups with doctors via video on the International Space Station (ISS)."

Meg: **"Yeah, they do show that in movies! There have been instances of violence due to isolation at Antarctic research stations. Do you know if there has ever been a documented mental break where someone became violent toward others in space?"**

Olivia Koski: "I think that you will be hard-pressed to find official documentation or anything on the record. I think there have been cases of sexual harassment of sorts. I wouldn't be surprised if there had been fights or even minor violence, but it probably would have been covered up. Again, I don't have any direct knowledge or official sources/documentation. Somehow the International Space Station doesn't seem that far away (it's only two hundred and fifty miles!), so it's almost less isolated than Antarctica. It would be interesting to compare the accounts (how isolated astronauts on ISS feel compared to Antarctic explorers). It seems like the environment on the ISS is more controlled and comfortable. And you're in constant communication with ground control. People are watching you constantly so if there hasn't been any violence, I would guess that is why. You're not all that isolated on the ISS versus on a longer term mission where communication is delayed and you don't have as much immediate contact with people on Earth."

After speaking with Olivia Koski, we got the impression that if there were violent dustups in space they would probably not be documented or spoken about! Ghosts, alcoholism, and other demons are explored in *The Shining*, but none as fascinating or iconic as the film's conveyance of extreme isolation. It doesn't take place in Antarctica or space, rather in a place normally busy and full of life. Perhaps that is why the Overlook Hotel, empty except for its dark past, is the true villain of *The Shining*.

CHAPTER TWENTY-ONE

THE RING

Year of Release: 2002	
Director: Gore Verbinski	
Writer: Ehren Kruger	
Starring: Naomi Watts, Martin Henderson	
Budget: $48 million	
Box Office: $249.3 million	

In 1991 the Japanese novel *Ringu* was published and became a phenomenon that inspired numerous films, video games, and a television series. The story focuses on Sadako, a girl who has been missing for decades, and a mysterious videotape that kills anyone who watches it. The 2002 adaptation changes the character name to Samara but the plot is very similar.

Ringu was based on the story of Okiku,[1] who died in a well outside of a castle in Japan. She was the servant to a samurai named Tessan Aoyama, and Aoyama took a particular liking to her. He fell in love with her but his feelings weren't reciprocated. In one version of the story, Okiku ended her own life by throwing herself down the castle's well, believing she had no other way out. In another version, Aoyama threw her down the well after she refused to be with him. In the wake of Okiku's death, she was said to crawl out of the well and appear to Aoyama on a nightly basis. Aoyama was driven insane by the vengeful ghost's screams in the night. Drawings of Okiku depict her as looking very similar to Sadako/Samara, with flowing black hair and a long white dress. This is the general depiction of a person who has died under unnatural circumstances in Japan: these ghosts are referred to as *Yūrei*, translating to either "faint soul" or "dim spirit." These tragic women are buried in white dresses, with their hair let down.

It is revealed in *The Ring* that you could survive in a well with only water for seven days, hence the phone call revealing when you will die after watching the cursed videotape. Is this scientifically accurate? According to an article in *Scientific American* an otherwise healthy individual has been known to live up to forty days without food.[2] There are cases though of overhydrated people dying within ten days.

Another aspect of *The Ring* is subliminal messages. When Rachel Keller (Naomi Watts) watches the deadly video, she is exposed to sudden, nearly imperceptible imagery. Could hidden messages cause us to act in a certain way? Subliminal influence can be traced back to the fifth century BCE. In ancient Greece, persuasive techniques were used in speeches to try to influence others.[3] The methods of logos, pathos, and ethos are still taught today in public speaking courses around the world. Aristotle believed that using these three "artistic proofs" could convince listeners to change their behavior, their thinking, or their actions. Logos is the appeal to logic. Speakers or advertisers use logos to state facts and give statistics. Pathos is the appeal to emotions. Imagery or stories that make us feel something use pathos. Ethos is the appeal to our ethics. Speakers will show their credibility and believability through ethos.

In 1943, the US government began embedding subliminal messages into mainstream advertising. "Buy Bonds" was one such message that the public was unknowingly inundated with to try to get them to help with the war effort. In 1957, a marketing expert faked a study that said popcorn and soda sales increased after he flashed the words "Eat Popcorn" and "Drink Coca-Cola" on the screen during movies. Although he admitted to doctoring the data, there do seem to be some studies that show that subliminal messages can make an impact on people's brains.[4] When subjects are viewing subliminal messages, there is a change in activity levels in various parts of the brain including the amygdala; where emotions are processed, the insula; the part of the brain that controls conscious awareness, the hippocampus; where memories are processed, and the visual cortex. Studies in the 1990s and 2000s showed that subliminal messages can have an effect on people's perception and can even affect academic performance. Several movies and TV shows explore the effects of subliminal messages and their possible negative effects. *A Clockwork Orange* (1971) shows a future in which aversion therapy is used to rehabilitate criminals. The famous, uncomfortable scene shows a man with his eyes held open being forced to watch violent imagery. In the

Subliminal messages have been prevalent in marketing for generations.

1988 movie *They Live,* subliminal messages are used to keep the general public subdued. People are being controlled by skull-faced aliens who only want humanity to "obey." Our favorite TV show, *The X-Files,* focuses on subliminal messages in the episode entitled "Blood." Townspeople are influenced by messages appearing on electronic devices that drive them to murder.

How are subliminal messages related to propaganda or brainwashing? Propaganda came more into the mainstream during World War II in the United States but has also been used for religious, political, and advertising purposes. The main goal of propagandists is to convince their audience to seek out information or begin to question certain beliefs. They use arguments that, while sometimes convincing, are not necessarily valid. Brainwashing takes this a step further and reduces the subject's ability to think critically for themselves. These techniques have been shown in classic books such as George Orwell's *1984* (1949) and J. R. R. Tolkien's *The Lord of the Rings* (1954). If we watch a mysterious videotape or video clip embedded with creepy, subliminal images will Samara crawl out of our screens and kill us? Hopefully not. But, becoming aware of subliminal messages and persuasive techniques can help us spot those who may want to do us harm.

SECTION EIGHT
FROM THE DEPTHS

CHAPTER TWENTY-TWO

CREATURE FROM THE BLACK LAGOON

Year of Release: 1954	
Director: Jack Arnold	
Writer: Harry Essex, Arthur A. Ross	
Starring: Richard Carlson, Julia Adams	
Budget: Unknown	
Box Office: $1.3 million	

A t a party one night during the filming of *Citizen Kane* (1941) Hollywood producer William Alland heard the tale of a half-human, half-fish creature. The story goes that an amphibious man would emerge from the Amazon River once a year to steal a woman from a local village. The woman would be taken back into the river and was never seen again. This story sparked Alland's imagination and in 1954 *Creature from the Black Lagoon* was released.

The story follows a group of scientists who are on an expedition to the Amazon. They find fossils that look human but with webbed hands and feet. Is it possible for humans to have webbed hands or feet? It is possible although extremely rare. Only one in three thousand children are born with this condition. How does it happen? While in the womb during the sixth or seventh week of pregnancy, a child's hands and feet begin to split and form fingers and toes. When this doesn't happen, the skin between fingers and toes remains combined together. Webbing can be associated with hereditary defects of both Down and Apert syndromes, which lead to unusual development of the bones in the hands and feet. Many people born with webbed fingers or toes can have surgery to separate them either at birth or later in life.

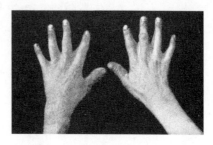

An example of webbed hands.

One character in the movie *Creature from the Black Lagoon* is studying lungfish, a fish that can breathe both on land and in water. Are there really animals that breathe air and live in water? Fish breathe air through their gills while marine mammals such as seals or whales have lungs and take in oxygen through their blowholes or snouts. Unlike humans, these marine mammals breathe voluntarily but similarly can stay underwater for about nine minutes. The human record for holding their breath the longest is German free diver Tom Sietas who held his breath underwater for twenty-two minutes and twenty-two seconds in 2012. Whales, dolphins, porpoises and some seals have a much greater ability to stay submerged and can remain there as long as ninety minutes. Turtles can hold their breath for about two hours. Other animals that can live in or out of the water include mudskippers, walking catfish, and climbing gourami. In 2013, there were reports of a Northern snakehead, a fish that can grow up to three feet in length, "walking" around Central Park in New York City. These terrifying creatures can survive up to four days on land and have no natural predators in the United States. They can reproduce at an incredible rate, laying tens of thousands of eggs multiple times per year. (Now that sounds like the start of a horror movie!)

Could examples like this be responsible for the numerous legends about sea creatures? There are stories of human-like beings across cultures that go back centuries. Mermaids are prevalent in lore ranging from Greek mythology to sightings from Europe, Asia, and Africa. Even Christopher Columbus was said to have spotted "three female forms which rose high out of the sea, but were not as beautiful as they are represented" in 1493.[1] In stories, mermaid-like creatures are responsible for shipwrecks, can be helpful, and even fall in love with humans. Adaro were malevolent merman-like sea spirits found in the mythology of the Solomon Islands. They were considered dangerous and arose from the wicked part of a person's spirit. An Adaro is described as a man with gills behind his ears, tail fins for feet, a horn like a shark's dorsal fin, and a swordfish-like spear growing out of his head.[2]

Dagon is a fish god that appears in the Hebrew Bible as well as ancient Sumerian texts. Dagon was worshipped and appears in popular fiction including this passage from *Paradise Lost* (1667) by John Milton:[3]

Dagon his name, sea-monster, upward man
And downward fish; yet had his temple high

H. P. Lovecraft also wrote of Dagon in his story by the same name and in *The Shadow over Innsmouth* (1931) A movie entitled *Dagon* (2002), based on his stories, features fish-human hybrids. Dagon also appears in the game *Dungeons and Dragons*.

Vodyanoy is a Slavic fairy tale character described as "a naked old man with a frog-like face, greenish beard, and long hair, with his body covered in algae and muck, usually covered in black fish scales. He has webbed paws instead of hands, a fish's tail, and eyes that burn like red-hot coals."[4] These creatures are thought to drown people who come into their territory, and store their souls.

Finfolk are also known to kidnap humans and keep them as slaves. Finfolk are part of Scottish folklore and were recorded as late as the nineteenth century.[5] The Loch Ness Monster is perhaps the most famous sea creature from this region. The first known sighting of it is from the sixth century. *The Life of St. Columba* by Adomnan chronicles his travels to Loch Ness. He writes about a man that had been eaten by "a monster" in the river and recounts his own face-to-face meeting. Later reports in the 1930s claim the Loch Ness Monster crossed roads and even emerged long enough to get its photo taken. Most recently a team of scientists, who are DNA experts, conducted two weeks of sample collecting in 2018 to determine what lives in the lake where "Nessie" has been spotted. No word yet on the results.[6]

There were reports of abductions in the late 1980s and early '90s that sounded strikingly similar to the *Creature from the Black Lagoon*. In "Abduction Notes," published in the April 1993 issue of the *MUFON UFO Journal*, hypnotherapist John Carpenter said:

> Typically, these reptilian creatures are reported to be about six to seven feet tall, upright, with lizard-like scales, greenish to brownish in color with claw-like, four-fingered webbed hands. Their faces are said to be a cross between a human and a snake, with a central ridge coming down from the top of the head to the snout. Adding to their serpent-like appearance are their eyes which have vertical slits in their pupils and golden irises.

The Gill-man in *Creature from the Black Lagoon* attacks because his home is being invaded by visitors. Are there instances in nature of animals fighting back against imposing forces? There are numerous examples all across the world of animals attacking humans when they encroach on their territory. Alligators in Florida, lions in Tanzania, and tigers in Bangladesh have all been reported in attacks. There are two main reasons for increases in these incidents. The first is a positive thing: conservation efforts. Because more species are being saved and more habitat is being set aside to preserve, the chances increase of human-animal encounters. On the flip side: as the human population increases globally, more land and food are being taken up. Animals will attack for food or to defend their territory, and these attacks can be devastating. One incident in 2002 saw a single elephant taking the lives of twelve people in Nepal.[7] Experts say that although the number of people dying from animal attacks is increasing, it is still rare and lower than the number of those dying from disease, famine, and war.

If the Gill-man could have communicated that his territory was being invaded, would it have helped? In 2017's *The Shape of Water*, a movie inspired by *Creature from the Black Lagoon*, director Guillermo del Toro explores the romantic relationship between a creature and a woman. The character of Elisa Esposito (Sally Hawkins) is mute and communicates through sign language. She is able to connect with the creature in the movie by teaching him sign language, and they fall in love. Although the creature may be part human in *The Shape of Water*, there are many real-life examples of animals being able to communicate with humans. Perhaps the most famous case is Koko, a female western lowland gorilla that learned a modified version of American Sign Language. She was able to use more than one thousand signs and understand over two thousand spoken words.[8]

Gorillas aren't the only animals that have been known to communicate with humans. Dolphins, elephants, dogs, and birds have all been known to learn words or symbols when observed in studies. A border collie named Chaser learned over one thousand words and Kanzi, a bonobo, learned more than three thousand symbols.[9] What can we, or they, do with this knowledge? Because our brains are different than animal brains, we may never have the same things to converse about. We think about things differently. According to primatologist Joan Silk, "primates are endowed with cognitive abilities that are especially well suited to tracking social information. For example, primates are able to recognize individuals; identify kin; compute the value of resources and services; keep track of past interactions with group members; make transitive inferences; discriminate between cooperators and defectors; and assess the qualities of prospective rivals, mates and allies."[10] She suggests sticking to these topics if communicating with a primate.

Humans have explored less than 5 percent of the planet's oceans. What could be lurking beneath the surface? We may never know. But learning about the lore and legends that surround the deep and ways to better communicate with animals may help us if or when we run into our own creature from the black lagoon.

CHAPTER TWENTY-THREE

JAWS

Year of Release: 1975	
Director: Steven Spielberg	
Writer: Peter Benchley	
Starring: Roy Scheider, Richard Dreyfuss	
Budget: $9 million	
Box Office: $470.7 million	

Beachgoers are tormented by a man-eating great white shark in *Jaws*. Although the town of Amity in the film was fictional, movie fans distinctly remember avoiding the water after seeing the blockbuster in June of 1975. California native Cheri Gray Pierce recalls the feeling of the time. "I was just finishing up my sophomore year of high school and was spending a lot of time at the beach that summer. I remember a lot of people being freaked out about going in the water, hearing a lot of them breaking out in that 'shark music' while we were basking in the sun."

Based on Peter Benchley's 1974 novel, *Jaws*'s journey to the screen was not an easy one. The film quickly went over budget because of its ambitious creature designs. The animatronic sharks developed for the shoot malfunctioned so often that director Steven Spielberg chose to rely on suggestion, as well as composer John Williams's iconic score, to stir dread in viewers' hearts. This choice ultimately paid off, an example of the shadowy, less-is-more horror filmmaking that critics love. While the larger-than-life shark in *Jaws* is memorable, the film itself is a behemoth. Considered the first summer blockbuster, *Jaws* opened in over four hundred and fifty theaters after an expensive promotional campaign. Because of its massive success, this model of opening to a wide, summer audience became an American tradition. Actor and comedian Jody Kujawa recalls seeing the movie as a kid:

> I remember seeing *Jaws* as a child on television. My parents had the Peter Benchley book which I had read three times and spent hours looking at the cover. It was an absolute thrill to see this film for the first time, watching it come to life. But my greatest thrill happened this last summer when a local theater was showing it for a mere $5. I cleared a night and went to see it. If you ever get the chance to see it on a big screen, do it. It's a whole new experience. It's the way it was meant to be seen. After all, it created the summer blockbuster genre.

The jaws of a shark.

A real-life shark attack was the inspiration for *Jaws*. In New Jersey, four people were killed and one injured by shark attacks over twelve days in July of 1916. Although author Peter Benchley denies the connection, the incidents in New Jersey are mentioned in his novel. Parallel to the novel and movie plot, people were encouraged to keep going to the beaches even after the first attack. The State Fish Commissioner of Pennsylvania, in regard to the first victim, was quoted as saying:

> Despite the death of Charles Vansant and the report that two sharks having been caught in that vicinity recently, I do not believe there is any reason why people should hesitate to go in swimming at the beaches for fear of man-eaters. The information in regard to the sharks is indefinite and I hardly believe that Vansant was bitten by a man-eater. Vansant was in the surf playing with a dog and it may be that a small shark had drifted in at high water, and was marooned by the tide. Being unable to move quickly and without food, he had come in to bite the dog and snapped at the man in passing.

Four more attacks later occurred along the New Jersey shore, causing panic that led to massive shark hunts and an emergence of the shark as a symbol of danger.

Prior to 1916, it wasn't believed that a shark would, or could, attack a human fatally. The *New York Times* reported that "the foremost authority on sharks in this country has doubted that any shark ever attacked a human being, and has published his doubts, but the recent cases have changed his view." Newspaper cartoonists across the country began using sharks as imagery for politicians, German U-boats, and even polio.

We are familiar with sharks and their attacks through the news, media, and our beloved "Shark Week." What are some creatures from the deep that may not be so recognizable to us? The giant squid is a deep ocean-dwelling squid that can grow up to forty-three feet long. Tales of these squid may have led to the legend of the Kraken, a sea monster who lives in the sea near Norway. Kraken have been featured in numerous films and stories including *Pirates of the Caribbean* (2003) and a Georges Méliès film entitled *Under the Seas* (1907). The footage used was of an octopus in a bathtub attacking a toy ship. Talk about practical effects! Even Alfred Tennyson wrote about the legendary Kraken in this sonnet:[1]

Below the thunders of the upper deep;
Far far beneath in the abysmal sea,
His ancient, dreamless, uninvaded sleep
The Kraken sleepeth: faintest sunlights flee
About his shadowy sides; above him swell
Huge sponges of millennial growth and height;
And far away into the sickly light,
From many a wondrous grot and secret cell
Unnumber'd and enormous polypi
Winnow with giant arms the slumbering green.
There hath he lain for ages, and will lie
Battening upon huge seaworms in his sleep,
Until the latter fire shall heat the deep;
Then once by man and angels to be seen,
In roaring he shall rise and on the surface die.

Another frightening, actual creature in the sea is the goblin shark. It is thought that these sharks can trace their lineage back 125 million years and are sometimes called "living fossils." These sharks got their name from a translation of its old Japanese name *tenguzame*. A *tengu* is a Japanese mythical creature often depicted with a long nose and red face. They may not be pretty but since they live so deep in the ocean, it's very uncommon for humans to have interactions with them.

The black swallower sounds like it was created for a horror movie but is indeed real. It grows to be about ten inches long, but don't let its small size fool you. This fish is known for its ability to swallow fish larger than itself. Black swallowers have been found to have eaten fish so big that they can't be digested faster than they're decomposing.

When imagining a creature called the "vampire squid" it seems natural that it would be a blood sucker. Thankfully, it's not. The squid got its name from its cloak-like webbing, dark color, and red eyes. Like vampires, they prefer the dark. They live in the deep sea and when disturbed are able to release a bioluminescent mucus to stun predators.

Giant oarfish may be responsible for many sea serpent legends. These large fish are shaped like ribbons and considered to be the world's longest bony fish. Sightings have been reported measuring them as long as fifty-six feet and a confirmed weight of six hundred pounds. Even though giant oarfish have been around for centuries, the first reliable footage of one in its natural habitat wasn't captured until 2010. No need to worry about this fish taking a bite out of you because it has no teeth.

Japanese spider crabs have the longest leg span of any arthropod. Arthropods are characterized by their jointed limbs and these crabs can reach eighteen feet from claw to claw. They can weigh up to forty-two pounds and are considered a delicacy to eat. They can live up to one hundred years and even though no human attacks have been reported, they have definitely haunted some dreams.

Whether it be through classic horror movies like *Jaws* or new ones like *47 Meters Down* (2017) the public remains enthralled with sharks and shark attacks. The threat may be unlikely to happen to us but it is real and that makes it all the more terrifying.

CHAPTER TWENTY-FOUR

ALIEN

Year of Release: 1979	
Director: Ridley Scott	
Writer: Dan O'Bannon	
Starring: Sigourney Weaver, Tom Skerritt	
Budget: $9 million	
Box Office: $104.9 million	

lien is set in the year 2122. The writer, Dan O'Bannon, found inspiration from many other classic horror and science-fiction movies including *The Thing from Another World* (1951) and *Planet of the Vampires* (1965) while writing the script. The film was pitched as "*Jaws* in space" and was also described as "*The Texas Chainsaw Massacre* of science fiction." That was a lot to live up to! Did they deliver on those promises? Audiences would say yes. The movie was a box-office success and inspired numerous sequels, books, games, and toys. The movie was also critically acclaimed, receiving many positive reviews. Roger Ebert noted an interesting point about the cast of the movie in his review:

None of them were particularly young. Tom Skerritt, the captain, was forty-six, Hurt was thirty-nine but looked older, Holm was forty-eight, Harry Dean Stanton was fifty-three, Yaphet Kotto was forty-two, and only Veronica Cartwright at thirty and Weaver at twenty-nine were in the age range of the usual thriller cast. Many recent action pictures have improbably young actors cast as key roles or sidekicks, but by skewing older, *Alien* achieves a certain texture without even making a point of it: These are not adventurers but workers, hired by a company to return twenty million tons of ore to Earth.[1]

The tagline of the movie *Alien* is "in space no one can hear you scream." Is it true? Sound travels in waves like light or heat does, but unlike light or heat, sound travels by making molecules vibrate. In order for sound to travel, there has to be something with molecules for it to travel through. On Earth, sound travels to your ears by vibrating air molecules. In space there are no molecules to vibrate so there is no sound there—so no one would hear you scream!

Space travel as depicted in *Alien* may not yet be possible, but when it is, how would we prepare for it? In the 2017 book *Vacation Guide to the Solar System: Science for the Savvy Space Traveler!* authors Olivia Koski and Jana Grcevich explore what travel to space would be like. They argue that although humans haven't set foot on another planet *yet* it's very possible that scientists will find a way for the human body to withstand extreme radiation, long journeys, and difficult conditions. Preparing for space travel would, and does, require some very specific training. Anyone going into space would need to be medically fit, including having good vision and a healthy blood pressure in order to avoid problems in space. Astronauts now, and presumably in the future, go through training in how to work and move in microgravity.[2]

What does living in space do to the human body? Weightlessness causes muscles to atrophy and the skeleton to deteriorate. Lack of gravity causes fluid distribution changes. Initially those traveling in space get "moon face" from the rush of fluids to their upper half and can have changes in vision, smell, and balance. Eventually, fluids redistribute and astronauts regain their natural look and functions. Another common ailment from space travel is called space adaptation syndrome. It's similar to motion sickness and can cause nausea, vomiting, lethargy, and headaches. It typically lasts about seventy-two hours while the body adjusts.

There is so much we don't know yet about space. To learn more about some of the theories being studied, we interviewed Allen Lipke, a former science teacher who worked at a lab that studied dark matter and neutrinos. Sometimes we can learn about space from things here on Earth.

Kelly: **"I know you were a science teacher but can you tell us a little bit about the lab you worked in?"**

Allen Lipke: "Certainly. The neutrino lab is an abandoned iron ore mine that is about half a mile below the surface. The lab started in 1982 and they were doing research on the possibility of protons decaying."

Meg: **"What did they discover through that research?"**

Allen Lipke: "That was of interest to the particle physics community of that time but no one ever found one to disintegrate. They're more stable than what we expected them to be."

Kelly: **"How does that relate to neutrons?"**

Allen Lipke: "Neutrons, kind of the sister to the proton, is very unstable in the sense that neutrons not found in the nucleus of an atom have a half-life of about fifteen minutes. So, in fifteen minutes a neutron is going to decay into a proton, an electron, and something else. That something else is a neutrino."

Meg: **"That's why it's called the neutrino lab!"**

Allen Lipke: "That understanding goes way back to about 1930."

Kelly: **"I knew I should have paid more attention in science!"**

Allen Lipke: "Wolfgang Pauli proposed a theory dealing with the idea that we're missing something in this research. There's something we're not seeing. It took until the 1950s to actually confirm that neutrinos existed."

Meg: **"So it took twenty years to get that research going?"**

Allen Lipke: "Well, then they fell out of interest entirely until the 1990s."

Meg: **"Is that when the underground lab in Northern Minnesota came about?"**

Allen Lipke: "In the late 1990s a lab was proposed to be put into the Soudan Underground mine but in order to do so they had to build a chamber that would be in alignment with Formulab down by Chicago."

Kelly: **"Why did it need to align with another lab?"**

The Soudan Underground Lab.

Allen Lipke: "There was a beam that was projected from Formulab underground through the surface of the earth. At a point in Wisconsin they estimated the beam was about six miles below the surface of the Earth. Because the Earth is round, by the time the beam got to Soudan it would be right on the middle of the detector."

Meg: **"What did the detector consist of?"**

Allen Lipke: "It was made out of steel plates. They were hung kind of like big stop signs. They were twenty-five feet in diameter but in the shape of a stop sign. There was eleven million pounds of these steel plates put into position in the chamber. In between the steel plates we put in plastic."

Kelly: **"What was the plastic for?"**

Allen Lipke: "It's the means by which we would collect data from interactions between the neutrino and matter, the substance of the detector."

Meg: "What did you find?"

Allen Lipke: "We would get about two events per day. Two neutrinos per day would be detected coming from Chicago going through the earth. Trillions of neutrinos would pass from Chicago through the earth and go through our detector. Neutrinos are very, very small. The smallest particle in the universe. They hit almost nothing but occasionally one, two, three per day would hit it and that would be our data."

Kelly: "This makes me feel like there's so much more to learn and so much left to explore even here on Earth! Is that research still going on?"

Allen Lipke: "That went on through 2016. A new detector was built farther north in Minnesota near the Canadian border."

Meg: "How does this relate to dark matter and the research that the lab did?"

Allen Lipke: "Dark matter is an entirely different experiment. It started at about the same time that the neutrino experiment started. The current philosophy is that dark matter is a particle. Dark matter interacts with regular matter, the stuff that we think of being protons and neutrons and the elements of the periodic table. We think dark matter makes up about 25 percent of our universe. Dark energy makes up 70 percent of our universe. And the remaining stuff that we know something about makes up 5 percent."

Kelly: "We know nothing!"

Meg: "What's in that 5 percent that we know?"

Allen Lipke: "Most of it is hydrogen and helium but the periodic table makes up about .3 percent of the matter that we find in our universe."

Kelly: "How did they do experiments relating to dark matter if we know so little about it?"

Allen Lipke: "The experiment at Soudan was an attempt to see dark matter. It centered around germanium detectors that were in a chamber that was cooled down to .04 degrees Kelvin which is right next to absolute zero."

Kelly: "Sounds like a winter in Minnesota!"

Meg: "How did you get a chamber that cold?"

Allen Lipke: "We cooled it down to that temperature using liquid helium and then evaporated it very fast. That whole process is an engineering marvel in my mind all by itself. But we cooled it down to get the germanium crystals to the point where the atoms were barely moving. We had to think of the temperature as being nothing more than molecular motions. So, any little change is going to result in an event."

Kelly: "You said any little change, so did that cause any false results?"

Allen Lipke: "The elevator that brings people down to the twenty-seventh level below the surface ends up about one-hundred-and-fifty feet away from where the dark matter detector was located. All of a sudden, they noticed they were getting an event every so often, all day long. But then at night they stopped having these events."

Kelly: "Was it the elevator?"

Allen Lipke: "Yes. The vibrations coming through the rock and into the chamber were resulting in detection. Also, we were still getting quite a bit of cosmic radiation."

Meg: "Tell us more about that."

Allen Lipke: "Being at that level [half a mile underground] you and I would get about one hundred cosmic ray particles hitting us per day. On the surface we would get one hundred cosmic ray particles hitting us per second. So, a huge difference, obviously."

Kelly: **"Did that affect the data?"**

Allen Lipke: "The cosmic rays were hitting the detector, going through lead shielding, going through the safeguards to try to filter out junk and background noise. It wasn't sufficient to block it all. But yet the Soudan experiment continued and we ended up at the point where we were at our maximum."

Meg: **"Is that research still going on?"**

Allen Lipke: "That research was taken out around 2015. The continuation of that research is going up to Ontario, Canada. It's in a copper nickel mine that's still active about a mile and a half below the surface. It'll be deeper and with that greater depth will have less cosmic radiation."

Meg: **"So, what exactly is dark matter?"**

Allen Lipke: "That's exactly the question! We don't know. Dark matter is a particle, we think. We think it's a particle because it interacts with regular matter on a gravitational basis and so our galaxies are spinning faster than we expect them to spin given the amount of matter that we can see. Again, this goes back to the 1930s."

Kelly: **"So, scientists were thinking about this already back then?"**

Allen Lipke: Fritz Zwicky and Jan Oort were two theoretical physicists that threw out this theory but no one would give them any money to do research until probably 1985. The scientific community finally decided there is something out there and we don't really know what it is. Finally, they were able to get some money and do some experiments and research to prove what it is. That's why we're still sitting at a theoretical level.

Kelly: **"It's still fairly new!"**

Allen Lipke: "We haven't really been working on it long, maybe twenty years. There's a variety of experiments going on around the world now. Some of them are quite similar to each other. The one we did and the one in Ontario are the only ones like that. There's a lot of empirical, theoretical kind of evidence. For example, we can see the bending of light coming from outer space from other galaxies."

Meg: **"What's bending it?"**

Allen Lipke: "We don't see what's there to bend it. The Hubble telescope was designed to look at stars but it's stumbled across all kinds of things like that. There are a lot of theories out there. I just read one not so long ago saying the answer to the dark matter question is quasars. They create a lot of gravitational pull. We need those theories."

Kelly: **"What are some other theories you're interested in?"**

Allen Lipke: "Parallel universes. We don't really know if we're the only universe out there or if there's another universe adjacent to us. It's kind of scary!"

Kelly: **"It's fascinating!"**

Parallel universes or alternate dimensions have been explored in such movies as *Silent Hill* (2006), *In the Mouth of Madness* (1995), and *The Mist* (2007) as well as the TV shows *Stranger Things* (2016–present) and *Star Trek* (1966–1969). There are even theories about alternate timelines and time travel that have become popular in recent years. One thing we know for certain? There is a lot of space left to explore and scientific and technological advances will only make space travel more possible.

SECTION NINE
WITCHES

CHAPTER TWENTY-FIVE

THE WITCH

Year of Release: 2015	
Director: Robert Eggers	
Writer: Robert Eggers	
Starring: Anya Taylor-Joy, Ralph Ineson	
Budget: $4 million	
Box Office: $40.4 million	

he Witch (2015), set in 1630s New England, follows a family who are banished from their town over a religious dispute. One of the earliest records of a witch dates back to the Bible. The book of 1 Samuel, thought to be written between 931 BCE and 721 BCE, tells the story of King Saul looking for the Witch of Endor to summon the dead prophet Samuel's spirit to help him defeat the Philistine army. The witch raises Samuel, who then prophesies the death of Saul and his sons. The next day Saul's sons die in battle and Saul dies by suicide. Other Old Testament verses condemn witches, such as Exodus 22:18, which says, "thou shalt not suffer a witch to live." Additional Biblical passages caution against divination, chanting, or using witches to contact the dead.

Are "witch doctors" considered witches? The term witch doctor, a practitioner meant to protect and heal others from ailments presumed to be caused by witches, first came into use in 1718. The term later came to be used to describe African shamans in the 1800s and is sometimes still used to describe those who use non-traditional medicine in healing. The type of healing can include spiritual, natural remedies, and other holistic approaches to the human body and mind.

The publication of *Malleus Maleficarum*, written in 1486, is credited with first spreading witch hysteria. Known as the Hammer of Witches, *Malleus Maleficarum* labeled witchcraft as heresy and served as a guide to identify, hunt, and interrogate witches. For more than one hundred years *Malleus Maleficarum* was the highest selling book in Europe other than the Bible. Perhaps the most well-known case of witch trials took place in Salem, Massachusetts in 1692. More than 150 people were accused of witchcraft and eighteen were put to death. Numerous films and literature depict witch trials like the play *The Crucible* (1953) by Arthur Miller, the animated film *ParaNorman* (2012), and even the popular Disney film *Hocus Pocus* (1993).

What are some of the methods people used to "prove" someone was a witch? One method was called the "swimming test." As suggested in the first season of *Chilling Adventures of Sabrina* (2018–present), a suspected witch was thrown into a body of water, tied to a rock, and left to sink or float. If the person floated, they were certainly a witch. If they sank and drowned, they were human. This was derived from the "trial by water," an ancient practice where suspected criminals or sorcerers were thrown into rushing rivers to allow a higher power to decide their fate. This custom was banned in many European countries in the Middle Ages, only to reemerge in the seventeenth century as a witch trial that persisted in some locales well into the eighteenth century. Another variation of this test was the ducking stool. An accused witch would be strapped to a chair and ducked into a body of water as punishment or to prove she was a witch. Another test was to find the "witch's mark." As seen in *The Scarlet Letter* (1995) suspects were stripped and

An example of a ducking stool.

publicly examined for signs of a blemish that witches were thought to receive upon making their pact with Satan. It was believed that this "Devil's Mark" could change shape and color, and was numb and insensitive to pain. It was easy for even the most minor physical imperfections, such as scars or birthmarks, to be labeled as signs of being a witch. By not being submissive enough, women were often accused of witchcraft. In the trial of Rachel Clinton, her accusers made the case against her with the following: "Did she not show the character of an embittered, meddlesome, demanding woman; perhaps in short, the character of a witch? Did she not scold, rail, threaten, and fight?" Even the way someone looked could doom them to being accused. Reverend John Gaule in the 1640s insisted that "every old woman with a wrinkled face, a furr'd brow, a hairy lip, a gobber tooth, a squint eye, a squeaking voice, or a scolding tongue is not only suspected, but pronounced for a witch."

What about modern witches? Those who practice witchcraft are considered pagans. The word pagan stems from the Latin *pagini* or *paganus*, meaning "hearth" or "home dweller." In the 1450s the fear of witchcraft became more prevalent, and people began associating witchcraft and paganism with devil worship, evil hexes, and spells. Contemporary witches include those who identify as Wiccan and those who practice other forms of witchcraft. Wicca was introduced to the public in 1954 and became a recognized religion in the United States in 1986. What do Wiccans believe in regard to witchcraft? Many follow the belief that magic, as said by Aleister Crowley, is "the science and art of causing change to occur in conformity with will." Wiccans cast spells for healing, protection, or to banish negative energy. A central theme of belief is of the five elements: air, water, fire, earth, and spirit. These elements are invoked in rituals and often represented by a pentagram shape. The ethic of the Wiccan Rede, which first appeared in 1974 as a poem in *Earth Religion News*, is followed by many Wiccans and summarized by the last two lines:

> Bide the Wiccan Law Ye Must,
> In Perfect Love and Perfect Trust,
> Eight Words the Wiccan Rede Fulfill,
> An it Harm None, Do What Ye Will.

There are an estimated three million Wiccans in the United States alone and it is considered the fastest growing religion in the country.

Today, women have reclaimed the word "witch" as a term of empow-erment. In response to the recent "Me Too" movement, some men compared being called out for sexual harassment to a witch hunt. Writer Lindy West responded with a piece in the *New York Times* entitled "Yes, This Is a Witch Hunt. I'm a Witch and I'm Hunting You." Author Joanna Malita-Król explores how the feminist movement identifies with witches as "the free, independent woman who lives on the edge of patriarchal culture, and the Pagan witch stereotype which largely agrees with the feminist interpretation."[1] People may not be being accused of witchcraft and burned at the stake in this day and age, but those who identify as witches or are seen as any type of "other" may still have a difficult road ahead. Hopefully humankind can learn from the mistakes of the past and prevent similar prejudices in the future.

CHAPTER TWENTY-SIX

CARRIE

Year of Release: 1976	
Director: Brian De Palma	
Writer: Lawrence D. Cohen	
Starring: Sissy Spacek, Piper Laurie	
Budget: $1.8 million	
Box Office: $33.8 million	

arrie is a story of firsts. It is the first novel by Stephen King, published in 1974. It is the first of his works to be adapted to film, and it was the cinematic debut for actor John Travolta. The film itself begins with a first. Carrie White (Sissy Spacek) endures a typical, teenage rite of passage. While for many girls their first menstrual period is awkward and uncomfortable, for Carrie it is horrific. In what has become one of the most iconic horror introductions in the last fifty years, Carrie's routine shower becomes a blood-soaked terror, not unlike that *other* famous shower scene. And her rising horror is only worsened by her fellow students' assault. Carrie is mocked and even pummeled with tampons. So how does this innocent, weak, seemingly powerless girl become a modern-day witch?

We first must ask; what are the powers that witches supposedly possess? Are there scientific explanations for them? Carrie comes to realize that with her transition into womanhood she has a new, uncommon trait: telekinesis, the ability to move things with one's mind. Numerous movies and TV shows feature characters with this ability including Eleven on *Stranger Things,* Jean Grey of *The X-Men* (2000), and Roald Dahl's beloved character *Matilda* (1988). Does telekinesis exist? The study of this supposed ability became popular in the late 1800s when objects would move as psychics and mediums contacted the dead. These occurrences proved to be hoaxes, but telekinesis gained attention again several decades later when a researcher at Duke University named J. B. Rhine tested the theory that people could change the outcome of rolling dice with their minds. The results of the experiments were enough to convince him and some others that telekinesis exists. Researchers since have been unable to duplicate his findings and therefore regard it as pseudoscience.

Another witch-related power that people are said to possess is the ability to "witch" water. Also called dowsing, the process begins with the diviner guiding a Y-shaped stick or two L-shaped ones. Water, minerals, and even gemstones are said to be found this way as the stick moves and points toward the target. This technique has been used by the military to try to locate weapons and tunnels during the Vietnam War and by regular citizens looking for a place to drill their well. The scientific community has attributed this technique to the ideomotor phenomenon. When this happens, the subject makes movements unconsciously. This phenomenon could also explain Ouija boards and other techniques used by mediums.

Henry Gross with a dowsing rod.

What about witches riding broomsticks? To understand this popular image, we must first look at how it came about. During the Middle Ages, plants were used to make ointments or "witches' salves" for sorcery and other activities. Hallucinogenic chemicals, called tropane alkaloids, made from a number of plants including deadly nightshade, henbane, mandrake, and jimsonweed were included in these witches' salves. According to an investigation into a witch in the year 1324: "In rifleing the closet of the ladie, they found a pipe of oyntment, wherewith she greased a staffe, upon which she ambled and galloped through thick and thin." Another investigation in the 1400s concluded: "But the vulgar believe, and the witches confess, that on certain days or nights they anoint a staff and ride on it to the appointed place or anoint themselves under the arms and in other hairy places." These salves, containing hallucinogens, caused vivid feelings of flying. Put all of these things together and you get the convention of witches riding broomsticks.

What would it take for a person to actually fly on a broomstick? To understand flight, we need to understand Newton's laws of motion. First, an object in motion tends to stay in motion or an object at rest tends to stay at rest. In order to get in the air, the witch would need to create thrust. The second law of motion states that the force gravity applies to the broomstick is equal to the mass of the broomstick, multiplied by the gravitational acceleration. The third law says that for every action there is an opposite and equal reaction. The amount of force created by magic to propel the broomstick forward will then, in turn, produce that much thrust to keep the broom in motion.

In several witch movies people are possessed by witches. What would the scientific explanation be for witches possessing people? The afflicted share certain unusual behaviors such as convulsions, fainting, changes in vocal and facial structure, and the loss of personality or memory. Before we had knowledge of conditions that could reasonably cause such behavior, these symptoms would be viewed as sinister or unnatural, leading to mass hysteria and public witch hunts such as the Salem Witch Trials. Hundreds of years later, though, scientists have suggested several explanations for these symptoms, the first of which is ergot poisoning or ergotism. Caused by the ingestion of fungi belonging to the genus *Claviceps*, it produces hallucinatory effects in the afflicted and can cause victims to suffer from vertigo, crawling sensations on the skin, extreme tingling, headaches, hallucinations, and seizure-like muscle contractions. Rye bread was a dietary staple in the colonies at the time, and historians have recognized that New England was unusually cool from 1690 through 1692. These were the perfect cool, damp conditions for the fungi to grow. Another medical explanation for unnusual behavior prominent during the Salem Witch Trials was an epidemic of bird-borne encephalitis lethargica. This is an inflammation of the brain spread by both insects and birds. Symptoms include fever, headaches, lethargy, double vision, abnormal eye movements, neck rigidity, behavioral changes, and tremors. Several girls suffered from these symptoms, and doctors at the time concluded that they suffered from being possessed by witchcraft.

Another power witches are said to possess is controlling things with their eyes or shooting lasers out of them. This power was seen in Roald Dahl's *The Witches* (1990). To understand the scientific basis for this, we interviewed laser scientist Olivia Koski.

Meg: **"How did you get interested in lasers?"**

Olivia Koski: "After college I got a job at a small research company that studies laser radar. I loved every aspect of it—doing these tabletop experiments every day, setting up optical systems, shaping light at your whim. It was so much fun to work in the lab. Probably the moment I became hooked was when I built my first laser from scratch. Lasers require very careful alignment, and it's often an all-or-nothing thing in terms of getting an output. You have to line up these mirrors in a laser cavity perfectly, and line up the energy source for the laser perfectly. Everything has to be just right. It was very frustrating the first time I built a laser. It didn't work for a long time. But I hung in there, and the first time I got 'first light' and created a laser beam was a very satisfying experience."

Kelly: **"What's the coolest thing you got to do in relation to laser technology?"**

Olivia Koski: "The lasers I was working with were very powerful. You had to be very careful or you could cause some serious injuries. I'm not going to lie and say I never accidentally burned a component or two. The smell of burning plastic was usually a signal you should cut system power immediately. Or make a slight tweak to realign the beam if you're feeling brave. I didn't do this, but I heard of some folks having fun dropping frozen hot dogs in a beam path to slice them in two."

Meg: **"How close are we to being able to shoot lasers out of our eyes?"**

Olivia Koski: "Well, the sharks with lasers things is actually completely feasible, so it would be fairly realistic for a villain or water witch (are there merwitches?) to train a fleet of dolphins to wear lasers on their heads. As for shooting lasers out of your eyes, sure. In principle you could rig up some glasses or a headpiece with lasers attached to them. But I definitely wouldn't recommend that, just because your eyes are very sensitive and it's generally best to keep lasers away from your eyes, even if your intention is to point them away from your eyes. But technology is getting better and better, and photonics are getting smaller and smaller. I'm just thinking now if it would ever be possible to embed a laser within a contact lens. It might be some day, but I'm not sure I would want to be the one to test it out. And the power level would probably not be nearly enough to cause any damage. I like this question, though!"

Kelly: **"Is there an experiment that readers can safely do at home regarding lasers?"**

Olivia Koski: "You really can't be too careful with lasers, so I might say 'no.' But . . . as long as you don't point them in your eyes or at airplanes, you can have a lot of fun with laser pointers, dollar store mirrors, and a fog machine or even steam from a humidifier for example."

Meg: **"Sounds like a fun Saturday night!"**

Another power witches are said to possess is the ability to talk to or connect with animals. The idea of a witch's "familiar" became popular during medieval times. Familiars were known to help witches in their practice of magic. Contemporary practitioners see pets, wildlife, or invisible spirit versions of familiars as their magical aids. Familiars have been portrayed in TV shows and movies about witches including Salem the cat in *Chilling Adventures of Sabrina* and rabbits in *The Witch*.

Some witches are known to have the power of clairvoyance. Does it exist? Clair meaning "clear" and voyance meaning "vision" is the alleged ability to gain information about an object, person, location, or physical event through an extra sense. Although scientists consider this ability to be pseudoscience, there are people who claim to have the ability. Some clairvoyants claim to see past events, the future, or other levels of perception that others cannot see. Throughout history those who claim to see the future have often been revered and respected. Nostradamus began publishing his predictions in 1550 and people still analyze and study them today. Edgar Cayce is a more contemporary example of a self-proclaimed clairvoyant. He was one of the founders of the New Age Movement in the early 1900s and influenced its teachings regarding karma, reincarnated souls, astrology, holistic medicine, and dream interpretation. Some famous clairvoyant characters in horror movies are Tangina from *Poltergeist*, Johnny Smith in *The Dead Zone* (1983), and Lorraine Warren in *The Conjuring* (2013) series.

Witches are known to use spells to make others act a certain way or fall in love with them. Are there such things as love potions? Mandrake, henbane, and verbena are ingredients used in love potions that can be traced back to Biblical times. Psychologists say there are scientific reasons we may feel attracted to or fall in love with someone quickly. It takes between ninety seconds and four minutes to feel an initial attraction, and this is first due to our observation of a person's body language, the tone and speed of their voice, and what they say. There are also attraction messages being sent through scent. Pheromones are produced by humans, animals, plants, and even bacteria for a variety of reasons. They can be used to collaborate or fight sexual attraction and can even lead to initial feelings of unexplained love.

Love is at the center of *Carrie*. Carrie's powers are otherworldly or witch-like but at the heart of the story is a terrorized girl who craves love from both her mother and her peers. This juxtaposition of the mundanity of being a teenager along with the heightened depiction of the supernatural is what propels *Carrie* into a haunting and memorable story. Many of us can relate to feeling shunned or bullied in high school. Nathan Payne, who directed *Carrie: The Musical* in 2019 said "*Carrie* stands the test of time and could be set in any era because its themes are universal. Everyone experiences being a teen differently, but bullying is so common and I think revenge fantasies are the dark secret that most people have at some point." Carrie is given the power to seek revenge. And it is this revenge, taken too far, that ultimately kills her.

At the end of the film, when Sue Snell (Amy Irving) visits Carrie's grave in a hazy dream, we share her conflicted feelings. As the audience we feel pity for Carrie for the trials she endures and also fear her for the destruction she is capable of. It is this complicated duality that sets Carrie White in the horror pantheon with the likes of poor, vengeful souls like Frankenstein's monster. Although Carrie's powers are witch-like, it is her humanity that endures.

CHAPTER TWENTY-SEVEN

THE BLAIR WITCH PROJECT

Year of Release: 1999
Directors: Daniel Myrick, Eduardo Sánchez
Writers: Daniel Myrick, Eduardo Sánchez
Starring: Heather Donahue, Michael C. Williams, Joshua Leonard
Budget: $60,000
Box Office: $248.6 million

To fully understand the phenomenon surrounding *The Blair Witch Project*, one simply needs to glance at the film's budget in comparison to the money it amassed. On the cusp of the twenty-first century, *The Blair Witch Project* exploded, sending droves of thrill-seekers to the theater. At the Internet's infancy, the filmmakers capitalized on a unique strategy that paid off in millions. The marketing was done in such a way as to make people believe the footage, depicting three budding filmmakers in search of the Blair Witch in the remote Maryland woods, was real. "Found footage" movies hadn't become popular yet and many people went into the theater expecting to watch the last harrowing days of three missing people. This was perhaps the first movie to use viral marketing through the Internet and other channels to reach a wide audience and get buzz going for a film. Missing persons posters were put on merchandise and even IMDB, the Internet movie database, listed the actors as "missing" and "presumed dead" for a time. A website for the film was created that looked to provide proof of the Blair Witch and the legends surrounding her. The unique (at that time) marketing and excellent word-of-mouth led *The Blair Witch Project* to become a tremendous box-office success.

We had the incredible opportunity to speak with Simon Barrett, the screenwriter of the 2016 movie *Blair Witch*, to find out more about his process of balancing creativity with science.

Kelly: **"What is it about horror that gets your creativity flowing?"**

Simon Barrett: "I tend to be attracted to specific stories and scenarios, a lot of which might fall into the horror genre. But the genre itself doesn't necessarily appeal to me more than any other."

Meg: **"When you're developing a supernatural story, how do you balance real world science and believability?"**

Simon Barrett: "The advantage of telling a supernatural tale is that you're already letting the viewer know that your story isn't intended to be real, so you have a fair amount of license to get creative with the story's relationship to actual reality. That said, I tend to be a fairly harsh or pedantic viewer, and if a film gets something obviously factually wrong, it does often take me out of the movie, and sometimes I find it really frustrating. I mean, if I'm watching a horror movie, and instead of being caught up in the suspense of the narrative I'm thinking something like, 'Really, did no one on set know how a defibrillator works?' then that film has failed, at least for me in that moment. Anything that we as viewers know to be wrong will remind us that the story we're watching isn't real. So, in terms of the science of a supernatural element itself, I think it just has to maintain some kind of internal logic, so that nothing takes the viewer out of the film or makes them doubt its feasibility. I don't need to believe every detail of how kryptonite or a transporter beam works, but I do need for their effects to be consistent so I'm not distracted from the story, basically."

Meg: "**I agree!**"

Simon Barrett: "If you're writing supernatural horror that has any kind of technological element, I think you just need to make sure that your science isn't distractingly impossible. I'm willing to entirely buy the science in many horror films as long as the story is good; I don't watch supernatural horror films for an education, normally, so if the film is fun and scary, that's most of what matters. So, my answer is, I try to avoid anything that would take any hypothetical viewer out of the movie, regardless of their expertise. Which sometimes means a fair amount of research to make sure I thoroughly understand any subject about which I'm writing, but often just means just trying to avoid doing anything obviously stupid."

Kelly: "**Did you do any research on witch legends when developing Blair Witch?**"

Simon Barrett: "I didn't research any real-world witch legends during the development of *Blair Witch*, because the existing *Blair Witch* mythology created by the directors, cast, and crew of the original film was so rich and complex. My goal was to just focus on that. Especially once you dig into later materials, like Ben Rock's accompanying fake documentaries and all the *Blair Witch* books that were published, of which there were at least a dozen in varying formats, there were already so many ideas to explore. So honestly, aside from some research into the actual history of Western Maryland to make sure my story additions were as realistic as possible, all of my *Blair Witch* research focused on thoroughly exploring the fictional world built by the original *Blair Witch Project* itself."

Meg: "**Why do you think the original *The Blair Witch Project* resonated with audiences? And how did you make certain to capture some of that same sensibility in *Blair Witch*?**"

Simon Barrett: "The genius of *The Blair Witch Project* is that, to this day, I think it's the most realistic found footage horror film. The experiment of the film's making, having the actors improvise their characters and performances while experiencing the staged events of the film, led to footage that always feels spontaneous and credible. Part of that isn't just the great acting, but the fact that the film was originally intended to be part of a larger whole, so it constantly hints at a deeper mythology than what you're seeing. That feels like real life, where actual history is endlessly deep and unknowable, and people can't just speak in expository monologues that fully explain it. *The Blair Witch Project* feels real because its filmmakers created a massive, complicated mythology, and then barely referenced it at all. The viewer is left feeling like they're missing most of the puzzle, but if they want to research more, the filmmakers also created these amazing ancillary materials for that, like the website and so on. I knew following *Paranormal Activity* (2009) and a glut of found footage films in the past two decades that we never had any chance of imitating the verisimilitude of *The Blair Witch Project*; that's the kind of cultural phenomenon that can really only happen once. But I did want to continue the tradition of revealing only part of a mythology, to create the sense of a deeper mystery. I tried to provide answers to some of the questions posed by the original *Blair Witch Project,* but I also wanted to create new questions for our audience, and new mysteries. Not fully understanding the horror is part of the uniqueness of *The Blair Witch Project*; the film has a lot of ambiguity to it, and fans of the film continue to debate various interpretations of its onscreen events. It was important to me that, while we were intentionally making a less realistic film, our movie's mythology held up to that level of scrutiny."

Meg: **"Tell us about how you developed the creature, or the witch. How did you describe it in the script? Was it based on any real or imagined creature that came before?"**

Simon Barrett: "The creature you're talking about, which was played by Breanna Watkins in the film, was originally referenced in a lot of various, ambiguous ways in the script; I believe I called it the "tall figure" a lot. Eventually, before production began, I had to put together a draft where all of the onscreen supernatural elements were referenced by exact names, just so that our crew would know which one I was talking about in any given scene. That's the problem with writing mysterious horror; it doesn't quite work on a script level. You need to let the crew know exactly what they need to create. So, from that point forward I referenced it as the "Ellie Kedward thing," basing it on the Blair Witch Dossier story that Ellie Kedward—who was not the original Blair Witch, of course, but someone who got caught up in the haunting and was later blamed for it—was tortured to death by being stretched, which I embellished quite a lot, I think. So, this kind of elongated body we're seeing in the woods is what's left of her. That's pretty clearly hinted at within the film, but of course we're also hinting at other possibilities of what that thing in the woods might be, other victims past and future, with other scenes. The figure has tree-like elements, and we see Ashley go through the early stages of what could be a similar transformation with what could be a tree root entering her skin, to cite another obvious example. Adam was bothered by that Ellie Kedward name being in the script in that context, as he felt Ellie Kedward was only a vague possibility for what we were seeing; he always just called it the creature, or whatever. And I made sure to not use Ellie Kedward as a character name on our production documents, so that name wouldn't be listed in the credits and spoil the mystery. But then I didn't come up with a new name for the creature either. I should have called her "The Shape" or something, it just didn't occur to me, so whoever created our end credits not unreasonably ended up labeling the tall figure in the woods as the Blair Witch, so I think Breanna got like, a "Blair Witch stunt performer" credit for one of her scenes or something explicit like that. Which made a lot of viewers justifiably annoyed, thinking we were trying to show them the actual Blair Witch. Whoops. I've become a lot more communicative about my intentionally ambiguous ideas like that since that experience."

Kelly: **"Regardless of the character name, she was terrifying!"**

The trio in *The Blair Witch Project* may have been influenced by evil forces when they got lost on their expedition, but what should you do if you lose your way in the woods? The National Forest Service urges hikers and campers to prepare in advance for the unexpected. They recommend you pack enough food and water for the activity you plan as well as a compass that you know how to use. These items may have come in handy for the travelers in *The Blair Witch Project*, but then again, the

You should always have a compass or GPS when hiking or camping.

witch may have corrupted them anyway. Ultimately a GPS is ideal for going into the woods but is not foolproof. Sometimes they don't receive a signal or the battery dies. As we saw in the sequel *Blair Witch* (2016), even a GPS can fail. Cell phones may also not work because of a lack of signal so they shouldn't be relied on too heavily. Knowing your terrain and studying accurate maps is important before taking off into the woods.

The National Forest Service also recommends wearing sturdy hiking boots, clothes that can be layered, and additional socks in case the ones you're wearing get wet. Other survival essentials like matches and tools should be packed just in case. One of the most important things you can do is tell someone where you are going and how long you plan to be gone. This way, if you don't return when you're supposed to, someone can come looking for you.

What should you do if you get lost or separated from your group? Experts agree that your most important tool is keeping a positive attitude. (Easier said than done when you hear the Blair Witch outside of your tent at night!) There is a method, based on an acronym that can help you if you get lost. Remember to STOP. S is for doing just that. Robert Koester, a search and rescue expert and author of *Lost Person Behavior*, says that when you realize you are lost, the first thing you should do is sit down.[1] Try to stay calm and stay in one place. You shouldn't continue to move until you complete the next steps. T is for think. Think about how you got to where you are and what landmarks you should be able to see. O is for observe. Get your compass and map out and determine directions based on where you are standing. It's recommended to stay on a trail if you can find one or to follow a drainage or stream downhill. If you don't have a compass or map you can determine directions by observing where the sun is in the sky or by noticing some other signs in nature. Moss tends to grow on the north side of trees while spider webs tend to be on the south. P is for plan. Come up with some possible plans, think them through, and then act on one of them. It's usually best to stay put though unless you're confident in your plan.

Some other self-rescue tips include taking time to rest, staying hydrated, and taking care of problems as they arise. Is there anything else the trio in *The Blair Witch Project* could have done to try to get out of the woods? Experts say it's important to stay dry, warm, and safe. The best way to do all of those things is to find or build a shelter. Getting lost is more common than people think. The Oregon-based Mountain Rescue Association, which has member teams in about twenty states, completes about three thousand rescue missions each year.

Being lost in the woods is a common theme in horror movies. The 2017 movie *The Ritual* follows a group of hikers who get lost in the wilderness in Scandinavia and run into more than they bargained for. Stephen King's novel *The Girl Who Loved Tom Gordon* (1999) focuses on a nine-year-old girl lost in the woods and how she copes with her seemingly inevitable fate. Aside from viral marketing, the brilliance of *The Blair Witch Project* and *Blair Witch* is in taking full advantage of the very human fear of being lost. The reality of being lost in the woods is horrific enough, but add in a witch hungry for your soul, and the stakes rise.

SECTION TEN

CREATURES

CHAPTER TWENTY-EIGHT

THE DESCENT

Year of Release: 2005	
Director: Neil Marshall	
Writer: Neil Marshall	
Starring: Shauna Macdonald, Natalie Mendoza	
Budget: $3.5 million	
Box Office: $57.1 million	

There are particular fears which seem to be universal. The fear of the dark is one we nearly all have succumbed to, especially as children when our imaginations ran wild. The dark, utilized in horror films and stories for ages, is really the fear of the unknown. In the dark anything seems possible. Claustrophobia, the fear of tight and confining places, is also common. There is something about being walled in, with little air or space to move, that has compelled many people to take the stairs rather than risk a faulty elevator.

It is no wonder then that British filmmaker Neil Marshall explored both of these fears, and threw in some ugly and vicious creatures, to great effect in his horror film *The Descent*. The story of a group of female friends spelunking in the Appalachian Mountains starts like a complicated, character-driven drama, but quickly turns into unabashed horror when the cave collapses. If being trapped underground isn't bad enough, the women of *The Descent* encounter beings beneath the earth. These albino creatures, who have adapted to their dark environment, begin to brutally kill the women. Sarah Carter (Shauna MacDonald) is the lone survivor. She eventually escapes through a hole in the cave, covered in blood and dirt from her horrific fight for survival. Interestingly, in the UK cut of the film, there is a bleaker ending. Sarah's escape is a dream, and she never gets out of the cave, or away from the strange, subterranean creatures. This ending was considered too depressing for American audiences.

Caves and underground cave systems exist all over the world. An ancient city in Turkey called Derinkuyu that dates back to the eighth century BCE spans more than eight levels going as deep as 260 feet with more than six hundred entrances to the surface and is large enough to have held over twenty thousand people at one time. The temperature underground maintained a steady fifty-five degrees Fahrenheit and was ideal for living conditions year-round. The city was rediscovered in 1963 and researchers found kitchens, bedrooms, bathrooms, food storage rooms, oil and wine presses, wells, weapons storage areas, churches, schools, tombs, and domestic animal stables.

Throughout the past two decades, underground systems have been found on several continents including South America, Africa, China, and in Europe. In Egypt, there are still unexplored caves and tunnels under the Giza Plateau, and in Wales a tunnel system was discovered that reaches fifteen miles in length. North America is the setting for *The Descent* and, like the movie, there are plenty of caves to explore in real life. (Hopefully without the creatures depicted in the movie!) The Mammoth Cave in Kentucky is the longest cave system in the world featuring over four hundred miles of surveyed passages. Evidence of human life in the cave dates back to the archaic period which runs from 8000 to 1000 BCE. Mammoth Cave serves as the setting for H. P. Lovecraft's story *The Beast in the Cave* (1918) and is referenced in Herman Melville's *Moby Dick* (1851).

What have people throughout history thought about creatures that live underground? There are stories and legends of creatures who live underground dating back centuries. Beings who dwell in mines or caves are known as knockers in many cultures. Some believe knockers cause cave-ins by knocking on the walls of a mine. Others think they knock to warn miners of impending danger. The legend of knockers began in England and Ireland and has been documented in fiction like Stephen King's *The Tommyknockers* (1987).

The troll is another creature that is said to live underground. In Norse and Scandinavian folklore, trolls are referred to as man-eaters and turn to stone when coming in contact with sunlight. They were often written about as being strong, evil, and not cordial with humans. Stories are told of trolls who caused hurricanes and avalanches and are described as having deformed bodies, claws, and fangs.

Gnomes are said to be underground dwellers who can move as easily through solid earth as humans move through air. Gnomes appeared in stories and legends dating back to the 1500s and were thought to be the protectors of mines and treasures. Currently, gnomes are almost synonymous with gardens but originally were portrayed as reluctant to interact with or be near humans.

Were there ever people discovered to be living underground? Throughout history many religions have believed in a subterranean realm inside the Earth. In Hinduism the idea of *Patala*, or that which is below the feet, is described as having many levels of life below us ruled by *nagas*. Recently people have been discovered living in underground dwellings for a variety of reasons. In Bucharest, Romania, there are hundreds of people living in a tunnel system due to homelessness and closed orphanages.

The writer and director of *The Descent* explained that the creatures, referred to as crawlers, were human. "They've evolved in this environment over thousands of years. They've adapted perfectly to thrive in the cave. They've lost their eyesight, they have acute hearing and smell and function perfectly in the pitch black. They're expert climbers, so they can go up any rock face and that is their world."[1]

There are over four hundred miles of caves to explore in the Mammoth Caves alone.

Aboveground, vision is useful for spotting predators, but in the dark world of caves, eyes become of little use. We talked to Bob Maki, a former biology teacher (and Kelly's dad!) to learn more about how animals adapt to their environments deep underground.

Kelly: "These humans in *The Descent* evolved and adapted to their surroundings. Do you think that's a realistic portrayal of what happens in nature?"

Bob Maki: "I think what happened was exactly right. Their skin tone was white . . ."

Meg: "And what would explain that?"

Bob Maki: "When you're in the sun your body produces melanin to darken your skin tone to protect you from the sun's rays."

Kelly: "And why did their eyes look the way they did?"

Bob Maki: "Their eyes changed over time too because there was no light. They became pretty much useless because it was dark all the time."

Kelly: "It's like bats, right?"

Bob Maki: "Yes! They use their sense of hearing and echolocation to navigate around."

Meg: "Did the creatures in *The Descent* have evolved ears?"

Bob Maki: "It appeared that their ears became larger so they could better hear things. This would be an adaptation to hear things better in the dark and their surroundings."

Kelly: "What other adaptations did you notice?"

Bob Maki: "Their teeth appeared to evolve into more animal-like teeth like a wolf or a bear. They needed those teeth to kill and eat the kinds of animals they were surviving off of in the cave."

Meg: "Or to eat humans!"

Bob Maki: "They also appeared to adapt to walk on all fours to make it through the low height of the caves. The way they moved made it seem like they adapted to move through the small spaces and use their limbs more like animal appendages."

Kelly: **"That definitely upped the creep factor for me."**

Meg: **"You never like when things or people move in strange ways in horror movies!"**

Kelly: **"It's scary!"**

Bob Maki: "They also seemed to have a moist, shiny, slippery skin. I don't know why they had it exactly but it must have helped them survive someway."

Kelly: **"What are these things technically speaking?"**

Bob Maki: "Creatures that only live in caves are known as troglobites and people who live in caves are called troglodytes."

Meg: **"What are some examples of animals who live in caves without light?"**

Bob Maki: "There's an amphibian called an olm that kind of looks like a snake."

Kelly: **"I'm creeped out already!"**

Bob Maki: "I was reading about it and it can survive up to ten years without eating."

Kelly: **"Do I dare ask what it eats?"**

Bob Maki: "It eats bugs or whatever other small animal it finds and swallows it whole. There are a lot of different animals that only live in caves like snakes, salamanders, and some fish."

Meg: **"Have they adapted, too, to live in that environment?"**

Bob Maki: "They have the adaptations I mentioned for the other creatures and they can have slower metabolisms too to conserve energy."

Kelly: **"The other term, troglodyte? That's just used in relation to people who live in caves?"**

Bob Maki: "Correct. Prehistoric 'cave men' are considered troglodytes and others who have lived in caves throughout history."

Kelly: **"I don't know about you, Meg, but after seeing this movie the last thing I'm planning to do is go cave diving let alone move into one."**

Meg: **"That's a big nope from me, too!"**

Could creatures like the ones in *The Descent* exist? In 1944 a mining inspector's report filed by an Inspector Glenn E. Barger claimed that fifteen miners had been eaten alive by a humanoid monster down in the mines of Dixonville, Pennsylvania. He claims to even have come face-to-face with one of the creatures himself but was able to get away unharmed. Another report from England reads as follows:

A tribe of subterranean creatures who surface on Cannock Chase to hunt for food could be behind a rash of "werewolf" and Bigfoot sightings near Stafford. And the mysterious beings could also be responsible for a string of pet disappearances, it has been claimed. West Midlands Ghost Club, our area's top paranormal investigation group, say they have been contacted by a number of shocked eye-witnesses who claim they have come to face to face with a "hairy, wolf-type creature" at the beauty spot. A scout leader and a local post man are amongst the "credible" witnesses to contact the club. Theories behind the sightings range from a crazed tramp to aliens. But now another paranormal expert has put forward the theory the sub-human beast is not a werewolf at all—but a Stone Age throwback. The investigator, who wishes to remain anonymous, told us: "Strange sightings in this area have been made over many years by civilians, military, police, ex-police and scout leaders on patrol. Some incidents have been reported and logged but others not—some people don't want to be classed as 'mad.' The strangest rumour has come from a senior local resident who believes the mysterious intruders to be subterranean," he told us. "The creatures have made their way to the surface via old earthworks to hunt, for example, local deer."

Are there examples of human adaptations to survive their environments like in *The Descent*? According to geneticists at Stanford, "All genetic mutations start out random, but those that are beneficial to an organism's success in their environment are directly selected for and quickly perpetuate throughout the population, providing a uniform, traceable signature."[2] When an environmental stress is constant and lasts for many generations, successful adaptation may develop through biological evolution. For example, humans adapted biologically to their climates. Those who live in tropical climates tended to be tall and lean in order to lose heat while those in arctic climates tended to be short and wide to conserve heat.[3]

Humans have also adapted their circadian rhythms to adapt to life throughout the world.[4] Body temperature, blood pressure, and sleep and wake cycles are time-based and change in a cyclical manner. Circadian rhythms have been able to change in people as they migrate to new parts of the globe. We experience disturbances in our circadian rhythms when we travel and feel jet lag as do people with schizophrenia, bipolar disorder, depression, seasonal affective disorder, and autism.

Another way we are able to adapt is through acclimatization. The human body can adjust to altitude differences when traveling in the mountains.

During acclimatization, over a few days to weeks, the body produces more red blood cells to counteract the lower oxygen saturation in blood in high altitudes. Full adaptation to high altitude is reached when the increase of red blood cells reaches a plateau and stops. Those who live in high altitudes, such as Tibet, have adaptations that make it easier for them to live in those conditions year-round. The changes specifically happen in oxygen respiration and blood circulation. When people who are not used to high altitudes travel in such areas, they may get altitude sickness which can consist of nausea, dizziness, and headaches. Those who have genetically adapted to altitude over several generations are part of the 2 percent of humans on Earth who can thrive under those conditions.

We are also able to adjust to pressure changes when diving deep underwater. Human adaptation to water can increase as dives or time in the water increase. When we are in water our bodies automatically trigger what's called the diving response. Our heart rate slows, our blood vessels constrict, and our spleens contract. Studies have shown that the Bajau, a group of people indigenous to parts of Indonesia, have genetically enlarged spleens which enable them to free dive to depths of up to 230 feet. They can also hold their breath longer than most people.

Species ability to adapt to their environments is a truly awe-inspiring element of biological science. It has helped humans and other animals thrive under the Darwinian concept of natural selection. Though, this concept in the hands of a horror filmmaker becomes less awesome and more terrifying. If there is anything scarier than being in the dark, it is sharing that dark space with a dangerous creature who can see you.

CHAPTER TWENTY-NINE

TREMORS

Year of Release: 1990	
Director: Ron Underwood	
Writer: Brent Maddock, S. S. Wilson	
Starring: Kevin Bacon, Fred Ward	
Budget: $11 million	
Box Office: $16.6 million	

On its release in 1990, *Tremors* was not a hit. It premiered in fifth place, performing well below industry predictions. Yet, *Tremors* sparked a franchise that includes five sequels, one prequel, and a television series. This could best be explained by the original film's humorous take on the genre, as well as its terrific reviews. Film critic James Berardinelli applauded *Tremors*; "horror-comedies often tread too far to one side or the other of that fine line; *Tremors* walks it like a tightrope."[1]

Set in the fictional town of Perfection, Nevada, *Tremors* is the story of two buddies, Val McKee (Kevin Bacon) and Earl Bassett (Fred Ward), who have grown bored of their small, desert community. As they are headed out of town they happen upon several peculiar scenes, what we later come to understand are the effects of a subterranean worm wreaking havoc on the citizens of Perfection. Things soon develop into a classic creature feature when the worms, hungry for humans, reach the surface.

Science is at the forefront of *Tremors* thanks to the addition of Rhonda LeBeck (Finn Carter). Rhonda, a graduate student, happens to be studying seismology, the study of earthquakes, in the area. There are unusual findings in her equipment because of the huge worms' thunderous movements.

The waves on her seismograph look similar to the undulating lines of an EKG machine, and by studying these, Rhonda can tell there is something beneath the desert floor.

The sandworms in *Tremors* are called graboids. Do sandworms exist? Sandworms permeate fiction and horror movies and are seen in many well-known franchises. The sandworms in *Dune* (1965) protect the coveted spice of melange in the fictional desert of Arrakis. They, like the graboids in *Tremors*, are big enough to swallow a vehicle whole. Sandworms in the movie *Beetlejuice* (1988) are large enough to ride and seem to prevent the dead from leaving their homes. In real life? Sandworms the size of those mentioned don't exist (thankfully). Worms are defined by having long, tube-like bodies with no limbs. These include microscopic worms that we can't see with the naked eye but also include a twenty-two-foot African earthworm and a 190-foot bootlace worm that lives in the sea.

Can creatures live underground like the graboids? We spoke to Allen Lipke, a former science teacher who worked at an underground lab to ask him what it's like being half a mile underground.

Kelly: **"How is life different that deep underground? Is there any water?"**

Allen Lipke: "The water seeps in here and there. Some of it is coming from above. It comes down the shaft. And other water is extremely salty."

Meg: **"Salty? Why?"**

Allen Lipke: "It's been underground for millions of years. It's the saltwater leftover from an inland sea. It's ancient, ancient water, and so there's bacteria there. Recently, I was sitting around with some friends and one of them brought up an article about multicellular nematodes existing in some of these underground mines that are feeding on the bacteria down there. That just blows my mind."

Meg: **"So, there *is* life underground?"**

Allen Lipke: "There are indications of life deep underground that you just wouldn't expect to have gotten there. The bacteria were there because water from the inland sea seeped down and carried bacteria with it. They're a lithotrophic type of bacteria that break down rock and get their energy from the molecules that they break down.

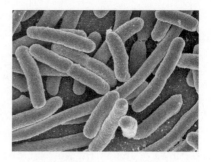

Bacteria can survive deep underground.

That leads to the idea of what will we find when we are able to do more research or look at Mars or the moons of Jupiter or the moons of Saturn. Some of these moons really look like they're possible for life to exist."

The article Mr. Lipke was referring to explains that rocks more than two miles underground were discovered to be over two hundred million years old, older than the dinosaurs, and were teeming with bacteria. These rocks provided proof that life can and does exist in places thought previously impossible to support life. Signs of life have been found everywhere from mines to the Arctic and under pressures and temperatures exceeding what were thought to be uninhabitable environments. The creatures of the deep are diverse and range from single-celled archaea to multicellular nematodes.[2] Nematodes can range in size from less than a millimeter long to as long as twenty-six feet when living inside of a sperm whale. Most species of nematodes have no effect on humans and their endeavors. They feed on bacteria, fungi, protozoans, and even other nematodes, and play a very important role in nutrient cycling and release of nutrients for plant growth. Other nematodes attack insects, and help to control insect pests.

What kinds of animals burrow underground like the graboids? Earthworms dig tunnels by ingesting the soil in front of them and excreting it with mucus to form burrow walls. Mammals, reptiles, amphibians, and even fish are known to burrow and make homes underground to use for protection from predators and climate or to store food. Burrows have been around for millions of years such as dinosaur burrows, which were recently discovered on multiple continents. They are not only dug into sand, however. Scabies mites burrow into human or animal skin to create their home while termites burrow into wood. These tunnels and holes can cause extensive damage to land and the environment as well as cause problems for humans.

Another aspect of the movie *Tremors* are the tremors themselves. What can make the ground shake? Earthquakes are one way we experience tremors of the earth. Naturally occurring earthquakes are explained by the moving of tectonic plates. When they shift, seismic activity is created and causes the ground to tremble. Earthquakes can also be caused by volcanic eruptions, mudslides, mining blasts, and nuclear tests. In ancient times people believed earthquakes were restless or angry gods. In Greek mythology Poseidon was also known as "earth shaker" and was credited with creating earthquakes with his giant trident. In Japan, people though a giant catfish was shaking the ground. Others believed gases were trying to escape the center of the earth or that there were volcanic eruptions underground.

In *Tremors: A Cold Day in Hell* (2018) the graboids have moved to the underground arctic. Are there subterranean creatures that can survive in the cold? Antarctica has already been known to teem with microscopic life. Tiny organisms dwell on the ice, live inside glaciers, and exist in a rich microbial ecosystem underneath the thick ice sheet, where no sunlight has been felt for millions of years. In 2017 scientists discovered an underground oasis in Antarctica that was being heated by an active volcano. On the surface the average temperature is around -4 degrees Fahrenheit while the temperature in the caves can reach 77 degrees. The researchers found DNA traces of life in the caves including small animals, moss, and algae. They also speculate that there could be undiscovered species living underground.

When we think of the dark and seemingly endless boundaries of our oceans, it is easy to imagine there are species we may not have discovered. Closer to home, beneath our very feet, it is frightening to think there could be similar creatures. In *Tremors*, there are obscure hints, like atypical seismic waves, that point to the "graboids" beneath. We can only hope that if a new species is forming in the dirt, we get a tremor of a clue.

CHAPTER THIRTY

THE WOLF MAN

Year of Release: 1941	
Director: George Waggner	
Writer: Curt Siodmak	
Starring: Lon Chaney Jr., Claude Rains	
Budget: $180,000	
Box Office: $2.4 million	

"Even a man who is pure in heart,
and says his prayers by night;
May become a wolf when the wolfbane blooms
and the autumn moon is bright."

This poem, recited throughout *The Wolf Man*, tells a tale within a few simple lines. No one is safe from becoming a wolf if the circumstances are right. This is a horror film that allows the audience to connect with its monster because he is a person, just like we are.

The Wolf Man is considered to have had a tremendous influence on how Hollywood has portrayed werewolves on screen ever since its release in 1941. The film reference book *1001 Movies You Must See before You Die* states that the film "still remains the most recognizable and most cherished version of the [werewolf] myth."[1] Not only did the film portray werewolves in an interesting way, it also allowed the audience to feel empathy for them. Perhaps the reason why has to do with the writer. Curt Siodmak was Jewish and left Germany during the rise of Nazism. In an interview with the Writers Guild of America in 1999, Siodmak said, "I am the wolf man. I was forced into a fate I didn't want." Movie historian Constantine Nasr observed that Siodmak saw this movie as "the story of an outsider whose destiny was cursed by forces he could not control." *The Wolf Man* premiered just three days after Pearl Harbor was attacked in 1941, but was able to be a box-office success despite the real-world atrocities taking place at the time.

Was *The Wolf Man* the first werewolf movie? It was not. In fact, it wasn't even the first werewolf movie released by Universal Studios. In 1935 *Werewolf of London* premiered but did not go on to be as popular as the later film. The concept of werewolves has been mentioned in legends and stories dating back to the first century AD by author Petronius and by Gervase of Tilbury in the 1100s. Werewolves were said in European folklore to bear certain physical traits in their human form including having a unibrow, curved fingernails, low-set ears, and a swinging stride. A Russian superstition suggests that a werewolf can be recognized by having bristles under their tongue.

What is the medical explanation for those who appear, physically, to be werewolves? Hypertrichosis is a condition that is characterized by an abnormal amount of hair growth over the body. Those who performed in circuses and "freak" shows likely had this condition. In the womb, humans are covered with a thin layer of hair called lanugo. Usually this hair is shed before birth but in some circumstances, for those with congenital hypertrichosis lanuginosa, the hair remains. There are other types of hypertrichosis, including acquired, which is due to side effects of medication. All types can be treated with hair removal products and services or by changing medications. The first recorded case of hypertrichosis is Pedro González in the 1500s. Despite living and acting as a nobleman, González and his children, who were also afflicted with the condition, were not considered fully human in the eyes of society. It is believed that his marriage may have partially inspired the fairy tale *Beauty and the Beast* (1740).

The only known case of someone being born with hypertrichosis in the United States was Alice Elizabeth Doherty. She was born in Minneapolis, Minnesota in 1887 with a layer of two-inch long, blonde hair all over her body. She appeared in side shows, as did others with this condition, including Fedor Adrianovich Jeftichew, also known as Jo-Jo the Dog-Faced Boy, and Julia Pastrana, also known as The Bearded Lady.

Another condition linked to werewolves is clinical lycanthropy. This is defined as a rare psychiatric syndrome in which the affected person is under the illusion that they can transform into, has transformed into, or is a non-human animal. A study was conducted in the McLean Hospital that stated:

Alice Elizabeth Doherty was born with hypertrichosis in 1887.

. . . we identified twelve cases of lycanthropy, ranging in duration from one day to thirteen years. The syndrome was generally associated with severe psychosis, but not with any specific psychiatric diagnosis or neurological findings, or with any particular outcome. As a rare but colourful presentation of psychosis, lycanthropy appears to have survived into modern times.[2]

Throughout history several serial killers claimed to have been wolves during their crimes including Gilles Garnier, Manuel Blanco Romasanta, Peter Stube, and Jean Grenier.

Werewolves are said to transform during a full moon. Does the moon actually have an effect on people or animals? This belief has been around for centuries. The word "lunatic" comes from the Roman moon goddess "Luna." In ancient Greece and Rome, philosophers thought that the water in our brains must be subject to the same tidal motions as the sea. This would explain strange behavior whenever the moon was particularly full or large in the sky. What do experts say? There is a lot of anecdotal evidence claiming an increase in emergency room visits, crime, and accidents when there is a full moon. The science to support this theory, though, is lacking. A 2005 study by Mayo Clinic researchers, reported in the journal *Psychiatric Services,* looked at how many patients checked into a psychiatric emergency department during the evening hours over several years. They found no statistical difference in the number of visits on the three nights surrounding full moons versus other nights. Regarding how a full moon may affect animals, a study in 2006 stated that:

> . . . the lunar cycle may affect hormonal changes early in phylogenesis (insects). In fish, the lunar clock influences reproduction and involves the hypothalamus-pituitary-gonadal axis. In birds, the daily variations in melatonin and corticosterone disappear during full-moon days. The lunar cycle also exerts effects on laboratory rats with regard to taste sensitivity and the ultrastructure of pineal gland cells.[3]

In many movies featuring werewolves, the person who has transformed has no memory of their exploits. How does science explain temporary memory loss? There are a few medical conditions that could affect memory as shown in these horror films. The first is transient global amnesia. People suffering from this condition will often not know where they are or how they got there. This type of amnesia is shown in werewolf media such as *An American Werewolf in London* (1981) and the TV show *Being Human* (2009–2013). The causes of this type of memory loss could also be explained by becoming a werewolf: change in blood flow and strenuous physical activity.

The main feature of werewolf stories is their ability to shapeshift. There are numerous examples of shapeshifters in literature and legends such as Skinwalkers in the Navajo culture and Sauron in *The Lord of the Rings* trilogy. Are there instances of shapeshifters in nature? Discovered in 1998, the mimic octopus can shapeshift and take on the form of other animals or objects. It's even able to mimic color and texture. In 2011, a study found that cuttlefish were able to mimic pictures or other visual cues to "blend in" with their surroundings.[4] They can mirror nearby plants or structures to evade predators. Pufferfish use their stretchable stomachs and their ability to quickly ingest huge amounts of water or air to puff themselves up to several times their normal size. Some species also have spines on their skin to make them even less edible. Another sea creature that can shapeshift is the deepstaria jelly. It relies on its entire body and its ability to change shape in order to catch its prey. At first glance it looks a bit like a floating blanket or a plastic bag. But when enveloping its victims, it cinches its bottom shut to "create a balloon of death."[5] It lives in deep water in complete darkness and was only recently captured on film in 2017.

Moving onto land, there are many shapeshifters in nature including the mutable rain frog, a type of frog that skips the tadpole stage and develops directly within their eggs. The frog was spotted in 2009 by researchers who took it to be photographed. When they initially saw it, it had a spiny texture to its skin. The next morning it was smooth. Eventually they realized the frog was able to morph into the texture that would protect it best in its environment. In 2015 the tentacled caterpillar was discovered in Peru. At first glance the caterpillar may appear to be just a twig or a branch on a tree but if disturbed, the caterpillar has the ability to shoot out four, white-tipped tentacles from its back. Chameleons can also be considered shapeshifters due to their ability to blend in with their surroundings. Chameleon skin has a layer which contains pigments, and under the layer are cells with guanine crystals. When the space between the guanine crystals shifts it changes the wavelength of light reflected off the crystals. This can be used for camouflage but also in social signaling and in reactions to temperature and other conditions.

Not only animals can shapeshift. Organisms and materials can mimic others as well. Slime mold are organisms that can live freely as single cells but can come together to form multicellular reproductive structures. When a mound or ball of slime mold is physically separated, it can find a way to reunite. Biologist John Bonner[6] contends that slime molds are "no more than a bag of amoebae encased in a thin slime sheath yet they manage to have various behaviors that are equal to those of animals who possess muscles and nerves with ganglia—that is, simple brains." He also states that these amoebae are the least understood by scientists so far and have a lot to be studied.

We may not have proof of werewolves among us but there are vast amounts of knowledge yet to be discovered in the animal kingdom. Through science we will be able to come to a better understanding of the life that surrounds us and perhaps be inspired to write the next horror movie based on a real-life creature.

A FINAL NOTE

Horror movies have moved us, terrified us, thrilled us, and made us question the world around us. Before I (Meg) walked by Kelly in her *X-Files* shirt, our love of horror often felt like a secret, a misunderstood aspect of our lives. It wasn't until our chance meeting so many years ago that we learned how horror and fandom can bring people together.

If you share in our nostalgia of picking out VHS tapes in the video store, we hope that you never stop seeking out the spooky. And if you're new to the genre, our hope is that this book has intrigued you to become one of us!

In an age when we can find horror at the click of a button, it is more important than ever to share in the films that made us who we are. So, rise from your graves, emerge from the darkened shadows, and show yourself—we'll see you in the horror section!

ACKNOWLEDGMENTS

We couldn't have written this book if we had never met. Thank you to *The X-Files* for bringing us together!

Thank you to Nicole Mele and everyone at Skyhorse.

Thank you to all of the experts we interviewed.

To our Rewinders, thanks for listening to the podcast!

Thank you to our families, including our own little monsters; Campbell and Vienna, Fox and Dexter.

And to the women of horror and science who came before us, thank you for illuminating the darkness—making our path far less scary.

ABOUT THE AUTHORS

Kelly Florence (left) and Meg Hafdahl

Meg Hafdahl is a horror and suspense author. Her fiction has appeared in anthologies such as *Eve's Requiem: Tales of Women, Mystery, and Horror* and *Eclectically Criminal*. Her work has been produced for audio by *The Wicked Library* and *The Lift,* and she is the author of two popular short story collections including *Twisted Reveries: Thirteen Tales of the Macabre*. Meg is also the author of the two novels; *Daughters of Darkness* and *Her Dark Inheritance* called "an intricate tale of betrayal, murder, and small-town intrigue" by *Horror Addicts* and "every bit as page turning as any King novel" by *RW Magazine*. Meg, also the cohost of the podcast *Horror Rewind*, lives in the snowy bluffs of Minnesota.

Kelly Florence is a communication instructor at Lake Superior College in Duluth, Minnesota, and is the creator and cohost of the *Horror Rewind* podcast as well as the producer and host of the *Be a Better Communicator* podcast. She received her BA in theatre at the University of Minnesota–Duluth and got her MA in communicating arts at the University of Wisconsin–Superior.

ENDNOTES

Chapter One: Halloween

1. Minutaglio, Ruth. (2018) "The Untold Story of the Real Person Who Inspired Halloween's Michael Myers." *Esquire.*
2. Hagerty, Barbara Bradley. (2017) "When Your Child Is a Psychopath." *The Atlantic.*
3. Heide, Kathleen M. (1992) "Why Kids Kill Parents." *Psychology Today.*
4. Goodwin, Christopher. (2011) "My Child, the Murderer." *The Guardian.*
5. Allen v. United States, 150 U.S. 551 (1893)
6. Hutchinson, Stefan. (2008) *Halloween: Nightdance.* Devil's Due Publishing.

Chapter Two: Child's Play

1. "High Court Upholds New Trial in 'Halloween II' Murder Case." (1989) *The LA Times.*
2. "James Bulger Case: Timeline of Key Quotations." (2010) *The Daily Telegraph.*

Chapter Three: A Nightmare on Elm Street

1. Marks, Craig. (2014) "Freddy Lives: An Oral History of *A Nightmare on Elm Street.*" *Vulture.*
2. "Deaths of Asians in Sleep Still a Mystery." (April 24, 1988) *LA Times.*
3. Madrigal, Alexis. (September 14, 2011) "The Dark Side of the Placebo Effect: When Intense Belief Kills." *The Atlantic.*
4. Olunu, Esther et al. (2018) "Sleep Paralysis, a Medical Condition with a Diverse Cultural Interpretation." *International Journal of Applied and Basic Medical Research* 8 (3), 137.
5. Hutson, Thommy. (2016) *Never Sleep Again: The Elm Street Legacy: The Making of Wes Craven's A Nightmare on Elm Street.* Permuted Press.

Chapter Four: Psycho

1. Leigh, Janet, with Christopher Nickens. (1995) *Psycho: Behind the Scenes of the Classic Thriller.* Harmony Press.
2. McNally, Kieran. (Jan 2007) "Schizophrenia as Split Personality/Jekyll and Hyde: The Origins of the Informal Usage in the English Language." *Journal of the History of the Behavioral Sciences* 43 (1) 69–79.

3. Stevenson, Robert Louis. (1886) *The Strange Case of Dr. Jekyll and Mr. Hyde*. New York: Black's Readers Service Company.

Chapter Five: The Texas Chainsaw Massacre
1. History.Com Editors. (2009) "Real-Life Psycho Ed Gein Dies." *History.com*.
2. Biography.Com Editors. (2014) "Ed Gein-Murderer-Biography." *Biography.com*.
3. Shorey, Eric. (September 2018) "Is the *Texas Chainsaw Massacre* Based on a True Story?" *Oxygen.com*.
4. Harrigan, Stephen. (July 2014) "A Double Date with Leatherface." *Texas Monthly* 42 (7), 52–56.
5. Ramsland, Katherine. (June 2016) "Secrets of Psycho." *Psychologytoday.com*.
6. Holmes, R. M., and S. T. Holmes. (2010). *Serial Murder* (3rd ed.). Thousand Oaks, CA: Sage. "The Texas Chain Saw Massacre (1974)." *The Numbers*.

Chapter Six: The Silence of the Lambs
1. Valdez, Maria G. (July 2013) "Thomas Harris, 'Silence of The Lambs' Author, Reveals Hannibal Lecter Was Inspired by Real Life Mexican Doctor." *Latin Times*.
2. Harvey, Oliver. (August 2013) "My Chilling Meeting with the Elegant Killer Doctor Who Inspired Lecter Character." *The Sun*.
3. Oleson, J. C. "The Devil Made Me Do It: The Criminological Theories of Hannibal Lecter, Part Three." *Journal of Criminal Justice and Popular Culture* 13 (2), 117–133.
4. Emery, Lea Rose. (January 2006) "11 Brilliant Serial Killers with Extremely High IQs." *Ranker.com*.
5. "What Different I.Q. Scores Mean." *Wilderdom.com*.
6. Bonn, Scott A. (March 2016) "Our Enduring Love Affair with Dr. Hannibal Lecter. *Psychologytoday.com*.
7. Salidie, Palmira. (December 2017) "Archaeological Evidence for Cannibalism in Prehistoric Western Europe: from Homo antecessor to the Bronze Age." *Journal of Archaeological Method & Theory*. 24, 1034–1071.
8. Bever, Lindsey. (2016) "Cannibalism: Survivor of the 1972 Andes Plane Crash Describes the 'Terrible' Decision He Had to Make to Stay Alive." *The Independent*.
9. Mufson, Beckett. (2018) "We Asked an Expert Why You Think Cannibalism Is Gross." *VICE*.

Chapter Seven: Dracula
1. Gerard, Emily. (1885) "Transylvanian Superstitions." *Nineteenth Century Magazine*, 128–144.
2. Lallanilla, Marc. (September 2017) "The Real Dracula: Vlad the Impaler." *Live Science*.

3. Stoker, Bram. (1897) *Dracula*. London: Archibald Constable and Company.
4. Michael, Eric. (2011) "A Natural History of Vampires." *Scientific American*.

Chapter Eight: Nosferatu

1. "Richard Chase: The Vampire of San Francisco Was as Ghoulish as It Sounds." (March 2018) *All That's Interesting*.
2. Le Page, M. (2018) "Wildlife Retreats to the Dark of Night." *New Scientist*. 238 (3183), 14.
3. Than, Ker. (2005) "Behind the Recent Spate of Vampire Bat Attacks." *Live Science*.
4. Sullivan, Rob. (2007) "The Dirt: Myths about Man-Eating Plants—Something to Chew On." *San Francisco Chronicle*.

Chapter Nine: Jennifer's Body

1. Stephens, Walter. (2002) *Demon Lovers: Witchcraft, Sex, and the Crisis of Belief*. University of Chicago Press.
2. "Homicide Trends in the United States, 1980–2008" United States Department of Justice. 2010.
3. Harrison, Marissa A., Erin A. Murphy, Lavina Y. Hoa, Thomas G. Bowers, and Claire V. Flaherty. (2015) "Female Serial Killers in the United States: Means, Motives, and Makings. *The Journal of Forensic Psychiatry & Psychology*, 26 (3), 383–406.
4. Geberth, Vernon J. (1995) "Psychopathic Sexual Sadists: The Psychology and Psychodynamics of Serial Killers. *Practical Homicide Investigation Law and Order*, 43 (4).
5. "Jolly Jane Topan: The Killer Nurse Obsessed with Death." (2018) *New England Historical Society*.
6. Mitchell, Dawn. (2017) "Female Indiana Serial Killer, the 'Comely' Belle Gunness, Loved Her Suitors to Death." *IndyStar*.

Chapter Ten: Night of the Living Dead

1. Rath, Arun. (July 2014) "The Secret Behind Romero's Scary Zombies: 'I Made Them the Neighbors.'" NPR.org.
2. Rosenfield, Kat. (2013) An Expert Explains *The Walking Dead* Timeline of Zombie Decay. http://www.mtv.com/news/2358856/walking-dead-zombie-decomposition/
3. Rath, Arun. (2014) "The Secret Behind Romero's Scary Zombies: 'I Made Them the Neighbors.'" NPR.org.

Chapter Eleven: Frankenstein

1. Shelley, Mary. (1818) *Frankenstein*. Lackington, Hughes, Harding, Mavor & Jones.
2. "Frankenstein." (March 1910) *Edison Kinetogram* 2, 3–4.
3. Calne, Roy. (2006) "History of Transplantation." *The Lancet* 368.

4.   Ho, Karl. (August 2002) "Seeing Dead People," *Straits Times/Asia News Network via Nation Weekend.*

Chapter Twelve: The Mummy

1.   Vieira, Mark A. (2003) *Hollywood Horror: From Gothic to Cosmic.* Harry N. Abrams.
2.   Osborne, Lawrence. (April 2001) "Grave Errors." *Lingua Franca: The Review of Academic Life.* 11 (3), 25.
3.   Ajileye, Ayodeji Blessing. (2018) "Human Embalming Techniques: A Review." *American Journal of Biomedical Sciences.* 10 (2), 82–95.
4.   "The Most Fascinating Ancient Burial Rituals." (March 18, 2016) *Talk Death.*
5.   Scovil, Lindsay. (2015) A Market for Death. www.amarketfordeath.com.
6.   Brenner, Erich. (2014) "Human Body Preservation—Old and New Techniques." *Journal of Anatomy* 224, 316–344.

Chapter Thirteen: The Exorcist

1.   Brinkley, Bill. (August 19, 1949) "Priest Freed Boy of Possession by Devil, Church Sources Say." *The Evening Star.*
2.   Allan, Thomas B. (2000) *Possessed: The True Story of an Exorcism.* iUniverse.
3.   Opasnick, Mark. (2005) "The Cold Hard Facts behind the Story That Inspired 'The Exorcist.'" *Strange Magazine.*
4.   *De Exorcismis et Supplicationibus Auibusdam.* The Vatican. 1999.
5.   Martin, Malachai. (1976) *Hostage to the Devil.* Harper One.
6.   Sherwood, Harriet. (March 30, 2018) "Vatican to Hold Exorcist Training Course after Rise in Possessions." *The Guardian.*
7.   National Safety Council. www.nsc.org

Chapter Fourteen: The Tingler

1.   Aguilera, María, and Gienah Díaz. (November 2014). "Niño de 4 años murió tras ser picado por ciempiés gigante." *El Tiempo*
2.   Zabs-Dkar, Tshogs-Drug-Ran-Grol. (1994) *The Life of Shabkar: The Autobiography of a Tibetan Yogin.* Albany: SUNY Press, 295.
3.   Nordqvist, Christopher. (Feb 2018) "What's to Know about Parasites?" *Medical News Today.*
4.   Dye, Column Lee. (July 6, 2011) "Shark Attack: First by Vicious, Small Species." *ABC News.*
5.   Cellania, Miss. (April 3, 2014) "8 Parasites That Create Zombie Animals." *Mental Floss.*
6.   Miller, Sara G. (January 25, 2017) "Doctors Remove 6-Foot-Long Tapeworm from Man's Gut." *Live Science.*
7.   "The Great Gatsby in 3D: Top 10 Movie Gimmicks." (Jan 12, 2011) *TIME.*

Chapter Fifteen: Get Out

1. Debruge, Peter. (2017) "Film Review: *Get Out*." *Variety*.
2. Orne, Martin T., and A. Gordon Hammer. "Hypnosis." *Encyclopedia Britannica*.
3. Kand, Erick. (2019) https://hypnosisevents.com/.
4. Kreskin. (1973) *The Amazing World of Kreskin*. Random House, 143.
5. Dwivedi, Sachin K., and Anuradha Kotnala. (December 2014) "Impact of Hypnotherapy in Mitigating the Symptoms of Depression." *Indian Journal of Positive Psychology* 5 (4), 456–460.
6. Thompson, Brian. "Recovered Trauma Memories and Hypnosis." https://www.mentalhelp.net/blogs/recovered-trauma-memories-and-hypnosis/
7. Thompson, Brian. "Recovered Trauma Memories and Hypnosis." https://www.mentalhelp.net/blogs/recovered-trauma-memories-and-hypnosis/
8. Coram, Gregory J., James L. Hafner (December 1988) "Early Recollections and Hypnosis." *Individual Psychology: The Journal of Adlerian Theory, Research & Practice* 44 (4), 472–480.
9. Andrews, Travis M. (November 15, 2016) "An Ohio Lawyer Hypnotized Six Female Clients and then Molested Them. Now He's Going to Prison." *The Washington Post*.

Chapter Sixteen: Cujo

1. Hagen, Elizabeth. (Oct 2009) "Why Bats Are Good." *ASU School of Life Sciences Ask a Biologist*. https://askabiologist.asu.edu/explore/bats
2. "What You Should Know about Cynophobia." https://www.healthline.com/health/cynophobia#symptoms
3. Coren, Stanley. (Sept 2012) "How Many Dogs Are There in the World?" *Psychology Today*.
4. Hendry, Annabella. (2014) "Black Dog." *Supernatural Magazine*.
5. Doyle, Sir Arthur Conan. (1902) *The Hound of the Baskervilles*. London. Chapter 14.
6. Brontë, Charlotte. (1847) *Jane Eyre*. London.
7. "Rabies." (2018) World Health Organization.
8. Berezow, Alex. (2019) "Woman Suffers Terrifying Death from Rabies after Puppy Bite." *American Council of Science and Health*.

Chapter Seventeen: Arachnophobia

1. McKechnie, Claire Charlotte. (Dec 2012) "Spiders, Horror, and Animal Others in Late Victorian Empire Fiction." *Journal of Victorian Culture* 17 (4), 505–516.
2. Buddle, Chris. (May 2015) "Why Are We So Afraid of Spiders?" *Independent*.
3. Kempley, Rita. (Jun 1990) "Arachnophobia." *The Washington Post*.

4. Berg, Ken. (June 1990) "Actor Found 'Arachnophobia' Co-Stars No Big Hairy Deal." *The Orlando Sentinel.*

5. Nag, Oishimaya Sen. (Aug 2017) "The Most Dangerous Animals of Amazon Rainforest."https://www.worldatlas.com/articles/the-most-dangerous-animals-of-the-amazon-rainforest.html

6. Richards, Stephanie. "Venomous Spider Bites in the United States." https://www.terminix.com/pest-control/spiders/bites/venomous-spider-bites/

7. Hall, G. (2001) "Avondale Spider, Our Hollywood Star!" Lincoln, New Zealand: Landcare Research.

8. Svetkey, Benjamin. (July 1990) "Wrangling Real Tarantulas in 'Arachnophobia' *Entertainment Weekly.*

9. McCarthy, Erin. (July 2015) "18 Creepy Facts about Arachnophobia." http://mentalfloss.com/article/66197/18-creepy-facts-about-arachnophobia

10. Geertds, Megan S. (March 2016) "(Un)Real Animals: Anthropomorphism and Early Learning about Animals." *Child Development Perspectives* 10 (1), 10–14.

11. Gard, Carolyn J. (Jan 1999) "Coping with the Fear of Fear." *Current Health* 2 (25), 22–24.

12. King, Neville J., Peter Muris, and Thomas H. Ollendick. (May 2005) "Childhood Fears and Phobias: Assessment and Treatment." *Childhood and Adolescent Mental Health* 10 (2), 50–56.

13. Sher, Daniel. "Anxiety and Paralysis." *Calm Clinic.* https://www.calmclinic.com/anxiety/paralysis

Chapter Eighteen: The Birds

1. "How Many Birds Are There on Earth?" http://birding-world.com/many-birds-earth/

2. Kizer, KW. (July 1994) "Domoic Acid Poisoning." *Western Journal of Medicine* 161 (1), 59–60.

3. "Domoic Acid Information and History." (May 2009) https://web.archive.org/web/20090513065206/http://www.cimwi.org/stranded_domoic.html

4. Jacobs, Julia. (Oct 7 2018) "Drunk Birds? How a Small Minnesota City Stumbled into the Spotlight." *The New York Times.*

5. Smith-Strickland, Kiona. (June 24 2015) "Crows May Be as Intelligent as Apes, Scientists Say." *The Sydney Morning Herald.*

6. Metcalfe, John. (Aug 10 2017) "When Crows Attack." *City Lab.*

Chapter Twenty: The Shining

1. Dockterman, Eliana. (Feb 18, 2014) "How to Avoid Cabin Fever during the Endless Winter." Time.com.

2. Rosenblatt, Paul C., Roxanne Marie Anderson, and Patricia A. Johnson. (June 1984) "The Meaning of Cabin Fever." *Journal of Social Psychology* 123 (1), 43–54.

3.   O'Neill, Natalie. (Oct 30, 2018) "Antarctica Scientist Stabbed Colleague for Spoiling Book Endings." *New York Post.*

Chapter Twenty-One: The Ring

1.   Addiss, Steven. (1986) *Japanese Ghosts and Demons.* George Braziller, Inc.
2.   Lieberson, Alan D. (November 8, 2004) "How Long Can a Person Survive without Food?" *Scientific American.*
3.   Mullen, Lincoln. (August 18, 2011) "How to Persuade—With Ethos, Pathos, or Logos?" *The Chronicle of Higher Education.*
4.   Stern, Victoria. (Sep/Oct 2015) "The Fall and Rise of Subliminal Messaging." *Scientific American Mind* 26 (5).

Chapter Twenty-Two: Creature from the Black Lagoon

1.   Walker, Sally M. (1999) *Manatees.* Minneapolis: Carolrhoda Books, 7.
2.   Knappert, Jan. (1992) *Pacific Mythology.* Acquarian/Thorson Harper Collins, 15.
3.   Milton, John. (2001) *Paradise Lost.* New York: Signet Classic.
4.   Rose, Carol. (2001) *Giants, Monsters, and Dragons.* W. W. Norton & Company.
5.   Towrie, S. (2018) "The Heritage of the Orkney Islands." Orkneyjar.
6.   Wharton, Jane. (2018) "Final Search for the Loch Ness Monster as Experts Comb the Waters for His DNA." *Metro News.*
7.   Sample, Ian. (May 17, 2006) "Beware of the Animals." *The Guardian.*
8.   Borenstein, Seth. (July 15, 2018) "Koko the Gorilla Used Smarts, Empathy to Help Change Views." *Associated Press.*
9.   Cummins, Denise D. (August 18, 2017) "Yes, We Can Communicate with Animals." *Scientific American.*
10.   Silk, Joan. (2007) "Social Components of Fitness in Primate Groups." *Science.*

Chapter Twenty-Three: Jaws

1.   Tennyson, Alfred Lord. (1830) "The Kraken."

Chapter Twenty-Four: Alien

1.   Ebert, Roger. (May 3, 2008) "Great Movies: Alien (1979)." *Chicago Sun-Times.*
2.   Koski, Olivia and Grcevich, Jana. (2017) *Vacation Guide to the Solar System: Science for the Savvy Space Traveler!* Penguin.

Chapter Twenty-Five: The Witch

1.   Anczyk, Adam, and Joanna Malita-Król. (July 1, 2017) "Women of Power: The Image of the Witch and Feminist Movements in Poland." *Pomegranate* 19 (2), 205.

Chapter Twenty-Seven: The Blair Witch Project

1. Koester, Robert. (2008) *Lost Person Behavior*. dbS Productions LLC.

Chapter Twenty-Eight: The Descent

1. Redfern, Nick. (2015) *The Bigfoot Book: The Encyclopedia of Sasquatch, Yeti and Cryptid Primates*. Visible Ink Press.
2. Brooks, Cassandra. (July 16, 2009) "Adaptation Is Key in Human Evolution." *Stanford News*.
3. Hlodan, Oksana. (June 1, 2010) "Evolution in Extreme Environments." *BioScience* 60 (6), 414–418.
4. Wolford, Brooke. (November 28, 2014) "Humans Adapted to Life at Different Latitudes by Tuning Their Circadian Clocks." *National Human Genome Research Institute*.

Chapter Twenty-Nine: Tremors

1. Berardinelli, James. Rotten Tomatoes. 2019.
2. Ravindran, Sandeep. (February 29, 2016) "Inner Earth Is Teeming with Exotic Forms of Life." *Smithsonian Magazine*.

Chapter Thirty: The Wolf Man

1. Schneider, Steven. (2003) *1001 Movies You Must See before You Die*. Quintessence Editions. Hauppauge, New York.
2. Keck, Paul E., et al. (February 1988) "Lycanthropy: Alive and Well in the Twentieth Century." *Psychol Med*. 18 (1), 113–120.
3. Zimecki M. (2006) "The Lunar Cycle: Effects on Human and Animal Behavior and Physiology." *Postepy Hig Med Dosw*.
4. Hanlon, Roger. (June 2011) "Shape-Shifting Cuttlefish Can Mimic Pictures." *National Geographic*.
5. Simon, Matt. November 13, 2015) "Absurd Creature of the Week." *Wired*.
6. MacPherson, Kitta. (January 21, 2010) "The 'Sultan of Slime': Biologist Continues to be Fascinated by Organisms after Nearly 70 Years of Study." *Princeton University*.

INDEX